An Introduction to Brown Dwarfs

From very-low-mass stars to super-Jupiters

Online at: https://doi.org/10.1088/2514-3433/ad757e

AAS Editor in Chief

Ethan Vishniac, Johns Hopkins University, Maryland, USA

About the program:

AAS-IOP Astronomy ebooks is the official book program of the American Astronomical Society (AAS) and aims to share in depth the most fascinating areas of astronomy, astrophysics, solar physics and planetary science. The program includes publications in the following topics:

GALAXIES AND
COSMOLOGY

INTERSTELLAR
MATTER AND THE
LOCAL UNIVERSE

STARS AND
STELLAR PHYSICS

EDUCATION,
OUTREACH,
AND HERITAGE

HIGH-ENERGY
PHENOMENA AND
FUNDAMENTAL
PHYSICS

THE SUN AND
THE HELIOSPHERE

THE SOLAR SYSTEM,
EXOPLANETS,
AND ASTROBIOLOGY

LABORATORY
ASTROPHYSICS,
INSTRUMENTATION,
SOFTWARE, AND DATA

Books in the program range in level from short introductory texts on fast-moving areas, graduate and upper-level undergraduate textbooks, research monographs, and practical handbooks.

For a complete list of published and forthcoming titles, please visit iopscience.org/books/aas.

About the American Astronomical Society

The American Astronomical Society (aas.org), established 1899, is the major organization of professional astronomers in North America. The membership (~7,000) also includes physicists, mathematicians, geologists, engineers, and others whose research interests lie within the broad spectrum of subjects now comprising the contemporary astronomical sciences. The mission of the Society is to enhance and share humanity's scientific understanding of the universe.

Editorial Advisory Board

An Introduction to Brown Dwarfs

From very-low-mass stars to super-Jupiters

John Gizis

Annie Jump Cannon Professor of Astronomy, Department of Physics and Astronomy, University of Delaware, 223 Sharp Lab, Newark, DE 19 716, USA

IOP Publishing, Bristol, UK

ISBN   978-0-7503-3387-0 (ebook)
ISBN   978-0-7503-3385-6 (print)
ISBN   978-0-7503-3388-7 (myPrint)
ISBN   978-0-7503-3386-3 (mobi)

DOI   10.1088/2514-3433/ad757e

Version: 20241201

AAS–IOP Astronomy
ISSN 2514-3433 (online)
ISSN 2515-141X (print)

British Library Cataloguing-in-Publication Data: A catalogue record for this book is available from the British Library.

Published by IOP Publishing, wholly owned by The Institute of Physics, London

IOP Publishing, No.2 The Distillery, Glassfields, Avon Street, Bristol, BS2 0GR, UK

US Office: IOP Publishing, Inc., 190 North Independence Mall West, Suite 601, Philadelphia, PA 19106, USA

To Evangelos and Frances Gizis, my parents, and Sheri Chinen Biesen,
my wife and best friend.

Contents

Preface   xi

Acknowledgements   xii

Author biography   xiii

1   Introduction   **1-1**

1.1   Stars and Brown Dwarfs   1-1

1.2   Definitions and Brown Dwarf Research Paradigms   1-6

1.3   Discovery of Brown Dwarfs   1-11

     References   1-15

2   The Basics   **2-1**

2.1   Fundamental Parameters   2-1

2.2   Astronomy Units: Magnitudes and Parsecs   2-5

2.3   Spectroscopy: Velocities, Surface Gravity and Metallicity   2-6

     References   2-11

3   Spectral Types   **3-1**

3.1   Spectral Typing Strategy   3-1

3.2   M and L dwarfs   3-2

3.3   T and Y Dwarfs   3-14

3.4   Luminosities and Effective Temperatures   3-18

     References   3-20

4   Photometry and Astrometry   **4-1**

4.1   Photometry   4-1

4.2   Astrometry   4-7

4.3   Sky Surveys   4-9

4.4   The Observational H–R Diagram   4-15

     References   4-18

5   Structure and Evolution   **5-1**

5.1   Elements of a Model   5-1

5.2   An Overview of Ultracool Dwarf Models   5-7

5.3   Key Physics: Clouds, Metallicity, and Initial Conditions   5-9
        5.3.1 The Surface Boundary Condition   5-9
        5.3.2 Metallicity and the Equation of State   5-13
        5.3.3 Initial Conditions   5-14
        References   5-15

6   Atmospheres   **6-1**
6.1   The Problem   6-1
6.2   Radiative Transfer   6-2
6.3   Key Physics   6-6
        6.3.1 Equilibrium Chemistry   6-6
        6.3.2 Condensate Clouds and Rainout Chemistry   6-7
        6.3.3 Mixing and Non-Equilibrium Chemistry   6-9
        6.3.4 Failure of the One-Dimensional Model   6-12
6.4   Clouds   6-13
6.5   Grid Models   6-14
        References   6-17

7   The Solar Neighborhood   **7-1**
7.1   Brown Dwarfs Within 20 Parsecs and the IMF   7-1
7.2   Very Young Stars and Nearby Moving Groups   7-8
        References   7-10

8   Conclusions   **8-1**
8.1   Binaries as Tests of Models   8-1
8.2   Final Thoughts   8-4
        References   8-5

Preface

I hope that this book will help students get started in the study of brown dwarfs. The study of brown dwarfs is based on traditional astronomy observing techniques, but the underlying physics is a unique mix of planetary science and stellar astrophysics. This field has many wonderful advanced review articles but I have found them a difficult starting point. My intention is that the book will serve as a bridge from undergraduate astronomy and astrophysics to the professional literature. As such, it is not a replacement for the literature, and the book does not feature new compilations of data. I do, however, aim to make use of published data sets and open-source models.

Acknowledgements

I have been very fortunate in my career to work with so many great and supportive scientists: learning stellar evolution theory from Pierre Demarque as an undergraduate, being advised for my Ph.D. by I. Neill Reid, joining the 2MASS Rare Objects Team, and advising many wonderful undergraduate and graduate students, including Basmah Riaz, John Shaw, Phil Castro, Rishi Paudel, Dylan Hilligoss, Jinbaio Ji, and Easton Honaker.

France Allard, Adam Burgasser, Adam Burrows, Kelle Cruz, Roc Cutri, Trent Dupuy, Jackie Faherty, Suzanne Hawley, Davy Kirkpatrick, Sandy Leggett, Jim Liebert, Mike Liu, Mark Marley, Eduardo Martín, Dave Monet, Rebecca Oppenheimer, Neill Reid, Emily Rice, Andreas Schweitzer, Adam Showman, and Mike Skrutskie deserve special acknowledgment for many conversations over the years. Finally, I'd like to thank my astronomy colleagues at the University of Delaware—Judi, Sally, Jim, Fed, Stan, Harry, Vero, Barbara, and Dermott.

Author biography

John Evangelos Gizis

John Evangelos Gizis received his Ph.D. in Astronomy from the California Institute of Technology in 1998. He was a postdoctoral research associate at the University of Massachusetts at Amherst and a research scientist at the Infrared Processing and Analysis Center (IPAC) at the California Institute of Technology with the Two Micron All-Sky Survey (2MASS) before joining the Department of Physics and Astronomy at the University of Delaware. He is now the Annie Jump Cannon Professor of Astronomy at the University of Delaware. Dr. Gizis conducts observations of very-low-mass stars and brown dwarfs.

An Introduction to Brown Dwarfs
From very-low-mass stars to super-Jupiters
John Gizis

Chapter 1

Introduction

1.1 Stars and Brown Dwarfs

Our purpose in this book is to survey the field of brown dwarfs at an introductory level to enable students to begin research and aid their entry into reading the more advanced scientific literature. The study of brown dwarfs draws on lessons learned from stellar astrophysics and planetary science and complements the new field of exoplanets. While astronomers have discovered thousands of brown dwarfs over the last three decades, just as significant has been the vast expansion of large public data sets, numerical models, machine learning techniques, and access to computational resources. Today, any student can access accurate, detailed optical and infrared imaging of any point on the sky and thousands of photometric and spectroscopic measurements of cataloged brown dwarfs. On the theory side, we can access electronic tables with predictions by experts and also compute our own detailed models of the interiors or atmospheres of brown dwarfs using open-source numerical codes. Whenever possible, this book will use figures that any reader can reproduce using the accompanying Python notebooks.

> Python Throughout this book, we will use Python and the standard NumPy (Harris et al. 2020), Matplotlib (Hunter 2007), and SciPy (Virtanen et al. 2020) packages. Jupyter (https://jupyter.org/about) notebooks for all Python-based calculations and plots are shared at https://github.com/jgizis/Introduction-to-Brown-Dwarfs.

Let us begin with a survey of the stars and brown dwarfs in the immediate vicinity of the Earth to get some idea of their importance. Starting with the Solar System, we find a single star (The Sun, $1\,\mathcal{M}_\odot = 1.989 \times 10^{30}$ kg, $1\,\mathcal{L}_\odot = 3.828 \times 10^{33}$ erg s^{-1}). Its largest planet, Jupiter, is just 1/1000 the Sun's mass—more precisely, $\mathcal{M}_{\text{Jup}} = 1.899 \times 10^{27}$ kg $= 0.009\,55\,\mathcal{M}_\odot$. The nearest stellar system, α Cen AB, is easily visible to humans and consists of a binary pair of two stars at a distance of

1.33 pc (1 pc $= 3.086 \times 10^{16}$ m) with a much fainter red dwarf companion. This third star, Proxima Centauri, is just 1/6 the mass of the Sun and was only discovered in 1916 using telescopes and photographic plates. Next, the single red dwarf Barnard's Star is at a distance of 1.8 pc. The list continues with two more single low-mass red dwarfs too faint to be seen without a telescope, Wolf 359 at 2.4 pc and Lalande 21 185 at 2.5 pc. Finally, at 2.6 pc, we come to the brightest star in the sky, Sirius, a two solar mass star orbited by a faint white dwarf companion. Except for the white dwarf, all of these stars fit into the familiar OBAFGKM spectral classification scheme developed by Annie Jump Cannon, and all lie comfortably within a factor of a few of the Sun's mass, though we can certainly find more massive stars if we look to greater distances. These eight stars and one stellar remnant were all known by 1917, but in the twenty-first century, three new objects were added to this census.

With a mission goal of finding the nearest stars to the Sun, the NASA space telescope Wide-Field Infrared Explorer (WISE) was launched in 2009 and scanned the entire sky with infrared detectors. One of the half-billion stars in its catalog, WISE J104915.7-531906.1 or Luhman 16, proved to be a binary system just 2.0 pc away. To be sure, binary star systems are not unusual, and Luhman 16's orbit within the Galaxy is typical for "Population I" stars in the Galactic disk. The mass, however, of each component is just 3% of the Sun's mass and their luminosity is $\mathcal{L} \approx 2 \times 10^{-5} \mathcal{L}_\odot$. The low luminosity indicates that Luhman 16 A and B do not have nuclear fusion in their cores. Such objects are called **brown dwarfs**. The Luhman 16 brown dwarfs are roughly at the geometric mean between the Sun's mass and Jupiter's mass, suggesting that we might also compare them to gas giant planets. Jupiter's visible appearance is dominated by clouds, mainly ammonia, that form zones, belts, and storms. The Luhman 16 brown dwarfs also show clouds and weather, though they are far too hot for the clouds to be condensed ammonia. Another WISE discovery, WISE J085510.83-071442.5, is an apparently single object at 2.3 pc. It may be about 0.01 $\mathcal{M}_\odot$, perhaps even 0.003 $\mathcal{M}_\odot$, and is so cool it may have condensed water clouds. Many more brown dwarfs are known, indeed, by the time (2012–2014) these three nearest brown dwarfs were discovered, hundreds of more distant brown dwarfs had already been observed and classified into three new spectral types "L," "T," and "Y." It is a coincidence that the three new objects were one L dwarf (Luhman 16A), one T dwarf (Luhman 16B), and one Y dwarf (W0855-0714).

A widespread view is that brown dwarfs are more closely related to stars than planets in some important ways. We will call this view, and the research program associated with it, the STAR-LIKE paradigm. The concept of "star-like" requires a concept of ordinary stars, so we will first ask "what is a star?" Brown dwarfs certainly are classified as stars if we think of stars as fixed, luminous, unresolved point sources observed on the night sky (in contrast to the wandering solar system planets), and indeed, we will use that terminology throughout this book as we discuss observations. We will not hesitate to say that we measure the position and brightness of each star in an image, even if some of the "stars" are "brown dwarfs" or "white dwarfs." Still, we are aiming for a physics-based definition of stars, not a description of our pictures. An interesting early physical definition of stars was advanced by Eddington (1926) in *The Internal Constitution of the Stars*, having modeled how an outwards pressure gradient could balance gravity in spherically symmetric stars:

$$\frac{dP}{dr} = -\frac{G\mathcal{M}(r)}{4\pi r^2} \tag{1.1}$$

Eddington suggested that both radiation and (ideal) gas pressure should be considered, so that the physical definition of stars would be objects where both pressures are important:

> We can imagine a physicist on a cloud-bound planet who has never heard tell of the stars calculating the ratio of radiation pressure to gas pressure for a series of globes of gas of various sizes, starting, say, with a globe of mass 10 gram, then 100 gram, 1000 gram, and so on.

Eddington argued that only for gas spheres of mass 10^{33} g to 10^{35} g—0.5 $\mathcal{M}_\odot$ to 50 $\mathcal{M}_\odot$ are the gas pressure and radiation pressure comparable "where we may expect something interesting to happen." He identified these as the stars. He went on to speculate that radiation pressure might also play a key role in the formation of stars, so that lower-mass objects might continue to gain mass until radiation pressure stopped accretion and the objects became stars. However, the calculation of the ratio of radiation pressure to ideal gas pressure depends on the mean molecular weight— as Eddington noted, if the Sun is made of hydrogen then radiation would only be significant in the most massive stars. Cecilia Payne–Gaposchkin's 1925 discovery that the Sun is mostly made of hydrogen (Payne 1925) means that radiation pressure is rather unimportant in the Sun and lower-mass stars, making Eddington's clever definition of a star obsolete—instead the equality of radiation and gas pressure is taken as an estimate of the maximum mass of stars! That said, it is worth beginning with his definition because it reminds us of the importance of gas pressure in understanding stars, and that this approach was very productive even without a knowledge of the energy source.

A modern physical description of stars is that they contract until they reach the main sequence of the Hertzsprung–Russell (HR) diagram, where they fuse ("burn") hydrogen into helium until the supply of hydrogen in the core is exhausted—having this phase of hydrogen burning can be considered the defining trait of stars. Ultimately, after passing through various phases of stellar evolution and losing some of its mass, a stellar remnant such as a white dwarf, neutron star, or black hole, is left behind. The relative number of stars with different initial masses is the result of a much different physical process—the turbulent collapse and fragmentation of a molecular cloud. The distribution of stellar masses produced by this "star formation process" is called the **Initial Mass Function** (IMF). For masses greater than 1 $\mathcal{M}_\odot$, the IMF can be described by a steep power-law $dN/dM \propto M^{-2.35}$ (Salpeter 1955) favoring lower masses, and overall the majority of stars are M dwarfs less massive than $0.6\,\mathcal{M}_\odot$ with main sequence lifetimes longer than the current age of the Universe. How low in mass does the IMF go? Wolf 359 remained the lowest luminosity star known for several decades, although as an isolated object its mass could not be measured. George Van Biesbroeck took on the task of finding less luminous stars by searching for extremely faint companions to nearby stars using long photographic plate exposures. In 1944, he announced the discovery of a companion to the M dwarf Wolf 1055 that was three photovisual magnitudes less luminous than Wolf 359 (van Biesbroeck 1944). This companion star, now known as VB 10, remained the lowest known luminosity star until the 1980s. van Biesbroeck (1961)'s catalog of discoveries also included VB 8, a star intermediate between Wolf 359 and VB 10.

This observational work set the stage for the theoretical prediction of a lower-mass limit to hydrogen burning in stars and the existence of what we now call brown dwarfs. As Shiv Kumar recalled in 2003 (Kumar 2003), starting as a postdoctoral researcher in 1958, he took on the problem of extending models to treat stars below $0.10\,\mathcal{M}_\odot$. He realized that the ideal gas pressure law would break down under these conditions and that the stellar core would reach a maximum temperature as it becomes partially electron degenerated. As these stars continued to contract, the core temperature would decrease as they would evolve towards a state of complete electron degeneracy. Unfortunately, the journal papers he submitted in 1961 and 1962 were rejected because the referees did not believe the conclusions! We can, however, see more developed versions of his models in his American Astronomical Society Conference abstract (Kumar 1962) and later Astrophysical Journal paper (Kumar 1963a):

It is shown that there is a lower limit to the mass of a main-sequence star. The stars with mass less than this limit become completely degenerate stars or "black [brown] dwarfs" as a consequence of the gravitational contraction and therefore they never go through the normal stellar evolution.

Using the evolutionary models, we will compute in Chapter 5 with open-source software, Figure 1.1 shows a modern version of the core temperature as a function of time. The **hydrogen-burning limit**—$0.074\,\mathcal{M}_\odot$ in these models—divides the stars that settle into a constant core temperature with hydrogen burning and the brown dwarfs that instead cool. Meanwhile, in 1961, Chuiro Hayashi discovered that pre-main sequence stars are fully convective and contract vertically in the HR diagram along what are now called Hayashi tracks. Takenori Nakano (Hayashi & Nakano 1963) added the key physics to the model to treat lower masses:

It is found that stars on the zero-age main sequence have radiative cores for $M > 0.26\,\mathcal{M}_\odot$, but they are fully convective for $0.26\,\mathcal{M}_\odot \geqslant M \geqslant 0.08\,\mathcal{M}_\odot$. The stars less massive than $0.08\,\mathcal{M}_\odot$ are found to contract to configurations of high electron-degeneracy without hydrogen burning.

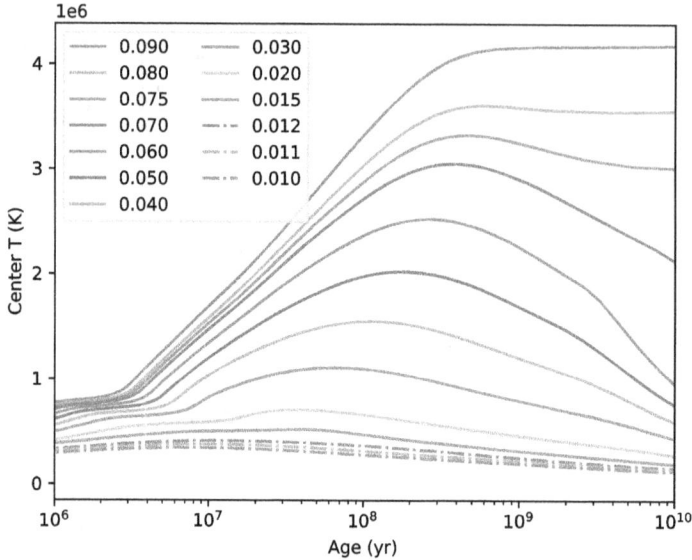

Figure 1.1. The central temperature of objects from 0.010 $\mathcal{M}_\odot$ to 0.090 $\mathcal{M}_\odot$ as a function of time as calculated in Chapter 5: The 0.070 $\mathcal{M}_\odot$ and lower-mass models reach a maximum central temperature and subsequently cool. Similar calculations by Shiv Kumar first predicted the existence of a lower mass limit to hydrogen burning and existence of what are now called brown dwarfs. There is also a deuterium-burning limit of ~13 $\mathcal{M}_{Jup}$ defined as objects that fully fuse their deuterium into helium. In these calculations, the 0.012 $\mathcal{M}_\odot$ (12.6 $\mathcal{M}_{Jup}$) model only fuses 40% of its deuterium.

As later recalled by Nakano (2014), these models were a major advance, representing "the first evolutionary path of brown dwarfs drawn on the HR diagram," and the determination of the hydrogen-burning limit agrees very well with modern calculations. Astrophysicists continued to develop the theory of very-low-mass stars and brown dwarfs over the following decades but the basic picture is present in Hayashi & Nakano (1963). Again using our own brown dwarf models, we show a modern theoretical HR diagram in Figure 1.2. The brown dwarf cooling tracks are closely spaced together on the HR diagram. Figure 1.3 shows the predicted luminosity as a function of time. Lower mass brown dwarfs cool to a given luminosity more quickly than more massive ones, and very-low-mass stars take ~10^9 yr to reach their main sequence luminosity. Other light isotopes (lithium, deuterium) may fuse but as Kumar noted in 1963 (Kumar 1963b), in much less time than the Kelvin–Helmholtz contraction timescale so they have little long-term importance for the evolution. Nevertheless, the pause in cooling due to deuterium fusion for ages is evident in Figure 1.3. The modern name "brown dwarfs" was coined by Jill Tarter (Tarter 1976, 2014) while investigating them as a candidate to explain baryonic dark matter. This gives us a first cut at physical definitions: The star formation process produces stars with masses above the hydrogen-burning limit that join the main sequence and also "brown dwarfs" with masses below the hydrogen-burning limit. We will, however, need to think more carefully about these definitions and what use they are.

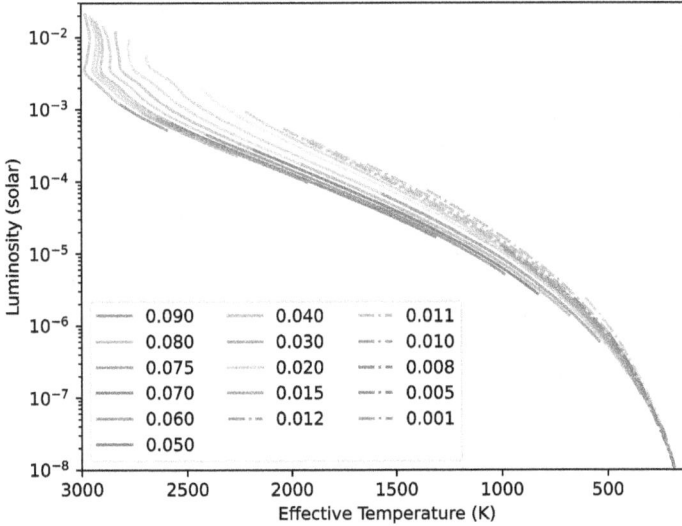

Figure 1.2. The evolution of objects with $0.010 \, \mathcal{M}_\odot$ to $0.090 \, \mathcal{M}_\odot$ in the theoretical H–R diagram as calculated in Chapter 5. Chuiro Hayashi and Takenori Nakano first calculated this evolution of brown dwarfs.

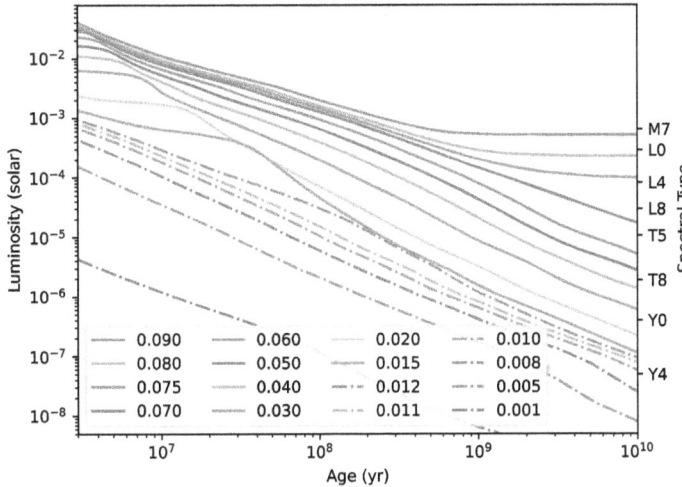

Figure 1.3. The evolution of objects with $0.010 \, \mathcal{M}_\odot$ to $0.090 \, \mathcal{M}_\odot$ as calculated in Chapter 5. An empirical relationship between spectral type and luminosity for ultracool dwarfs older than 200 Myr (Sanghi et al. 2023) is also shown.

1.2 Definitions and Brown Dwarf Research Paradigms

There is no doubt that countless objects with $0.100 \, \mathcal{M}_\odot > \mathcal{M} > 0.001 \, \mathcal{M}_\odot$ exist in the universe and they are worthy of study by astronomers regardless of what we call them. We want to observationally discover and characterize these objects and create and test theoretical models of their structure and formation. Some definitions should help us in this ambitious project. **Very-low-mass stars** for our purposes will be defined as stars with $\mathcal{M} \leqslant 0.10 \; \mathcal{M}_\odot$ but above the hydrogen-burning limit.[1] **Brown dwarfs** are sometimes defined as "objects below the hydrogen-burning limit but above the deuterium-burning limit" and sometimes as "objects below the hydrogen-burning limit that *form like stars.*" Both definitions avoid calling the planet Jupiter a brown dwarf! If we wish to specifically refer to what the second definition would call "brown dwarfs below the deuterium-burning limit," we will use the term **planetary-mass object**. (Others call these **planemos**, **free-floating planets**, **rogue planets**, or even simply **planets**. The suggestion of **sub-brown dwarf** has never been widely used.) As we know from the continued controversy over the International Astronomical Union (IAU) defining Pluto as a "dwarf planet" rather than a (major) planet, these definitions can seem arbitrary but are really disputes about concepts. Rather than quibble over different definitions of brown dwarf, we should emphasize instead how different concepts inform research paradigms that provide frameworks for asking and answering questions. Let us specifically consider the STAR-LIKE paradigm, the DARK-MATTER paradigm, the SUPER-JUPITER paradigm, and the BROWN-DWARF-TO-EXOPLANETS paradigm for brown dwarfs and planetary mass objects.

The STAR-LIKE research paradigm starts with the assumption that brown dwarfs should result from the star formation process. After all, if the turbulent fragmentation and collapse of a molecular cloud produces $10 \; \mathcal{M}_\odot$, $1 \; \mathcal{M}_\odot$, and $0.1 \; \mathcal{M}_\odot$ stars, why not also $0.05 \; \mathcal{M}_\odot$ stars? In a process that typically takes $< 10^6$ years, how could the hydrogen-burning timescale of 10^8 yr seen in Figure 1.1 be important? Furthermore, attempts to predict the theoretical minimum mass of fragmentation have long suggested that the minimum mass may be more like $0.005 - 0.01 \; \mathcal{M}_\odot$ (Low & Lynden-Bell 1976; Silk 1977), thereby predicting that the star formation process will create planetary-mass objects if these minimum mass objects do not subsequently accrete enough mass. Empirically, there is now no doubt that this assumption is true for many objects. The IMF observed in star-forming regions such as Taurus (Figure 1.4) shows that objects not just below the hydrogen burning but even the deuterium-burning limit are present. In the STAR-LIKE paradigm, we expect to see brown dwarfs everywhere we see stars: star-forming regions, open star clusters, the solar neighborhood, the Galactic (Population I) thin disk, the old metal-poor (Population II) halo, globular clusters, and so forth. They have hot starts as they collapse from proto-stellar fragments and release gravitational potential energy, and their elemental compositions should match the parent molecular cloud. Just as star formation creates binaries, triples, and higher order systems, brown dwarfs (including planetary mass objects) are expected to be companions to more massive stars and to have lower-mass companions of their own. It is not surprising that they have circumstellar disks and potentially would have exoplanets (or asteroids or comets). Some of these properties may be very mass dependent—we know the binary exoplanetary statistics are a function of primary

---

[1] In other contexts, very-low-mass stars may be defined as stars below the fully convective limit but above the hydrogen-burning limit. In any case, you should always avoid the temptation to call the Sun a "high-mass star" if you want other astronomers to understand you.

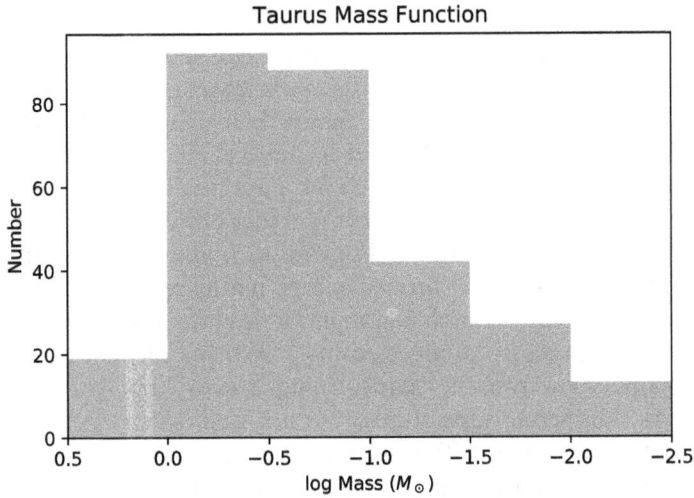

Figure 1.4. The observed Initial Mass Function (IMF) in the Taurus star-forming region (Esplin & Luhman 2019). The star formation process has resulted in stars, brown dwarfs, and planetary mass objects.

star mass, so we certainly need models and observations to study these issues. Key questions for this paradigm include: What is the IMF below 0.10 $\mathcal{M}_\odot$, how does it vary with environment, and what is the lower mass limit of star formation? Just as important as the questions are the tools that this paradigm provides us: We expect, for example, that the composition (or metallicity) of the brown dwarf population is similar to the stellar population, within the limits of evolutionary selection effects for both stars and brown dwarfs, so that we can study Population I and Population II brown dwarfs, depending on how the IMF varies. In this paradigm, we can use the composition and age of a primary star to infer the composition and age of a brown dwarf companion: note that in contrast, this does not work for the Solar System's planets and the Sun. For that matter, we can use the brown dwarfs to learn about the stars; the properties of very-low-mass stars and brown dwarfs are the most reliable way to age date nearby open clusters. The STAR-LIKE paradigm might also be called the FAILED-STARS paradigm, as brown dwarfs have often been described as failed stars, but we do not wish to introduce the negative connotation of failure into our field of study. After all, stars can be seen equally well as failed brown dwarfs that are temporarily prevented from reaching their final degenerate configuration by nuclear fusion.

The DARK-MATTER research paradigm suggested that low-luminosity brown dwarfs are a candidate form of baryonic dark matter, or MAssive Compact Halo Objects (MACHOs). Cosmologists imagined for this scenario that the different conditions in the early universe might have created a large number of brown dwarfs; for example, a metal-free Population III IMF might be shifted to very low masses. These hypothesized brown dwarfs would have been unobservable at the time, even though they would have to exist in very large numbers to be a significant contribution to the Galaxy's mass budget. Less exotically, we might also imagine that even in Population I, the total mass in brown dwarfs might be comparable to, even exceed, the mass in stars. This research paradigm is now finished, and the result is that brown dwarfs are not a significant dark matter component. Microlensing surveys showed that the Galaxy's dark matter halo is not made up of MACHO brown dwarfs or planetary mass objects. Indeed, the current Population III paradigm is that very massive stars ($>100\ \mathcal{M}_\odot$) are favored in zero-metallicity star formation. In any case, our modern surveys find that brown dwarfs contribute only a small percentage of the total mass for Population I, and would be able to directly detect nearby brown dwarfs that formed a dark matter halo.

The SUPER-JUPITER research paradigm brings attention to the likelihood that the "planet-forming process" produces a mass distribution that overlaps with the STAR-LIKE population from the "star-forming process." Unlike "super-Earths" or "mini-Neptunes," the term "super-Jupiter" is not well defined by the astronomical community, but often is used with respect to exoplanets that are close to or above the deuterium-burning limit, or at least many times the mass of Jupiter. Research topics in this paradigm are searching for objects that do not fit the STAR-LIKE paradigm. They might have very different "cold" initial conditions with low entropy and therefore different evolutionary tracks. They might be enhanced in metals, like Jupiter. They might be enriched or depleted in carbon or oxygen, having formed in proto-planetary disks where ices formed. All of these possibilities have been the subject of study for directly imaged "exoplanets" such as HR 8799bcde, 51 Eri b, and AF Lep b, but they can also be applied to isolated planetary mass objects. Remembering that we are more interested in research questions than arbitrary definitions, we can look for signs of different formation processes. Finding a bimodal IMF at planetary masses, for example, would be very interesting to test formation models. A recent study of the Orion star-forming region claims a new population of Jupiter-mass Binary Objects (JuMBOs) ($0.001\ \mathcal{M}_\odot \lesssim \mathcal{M} \lesssim 0.01\ \mathcal{M}_\odot$) —new in the sense that their properties are surprising in the STAR-LIKE paradigm. In July 2024, observations of the "super-Jupiter" ϵ Ind Ab showed it has a similar temperature to the coldest known brown dwarf, WISE J085510.83-071442.5, yet a very different spectral energy distribution (Matthews et al. 2024).

The BROWN-DWARF-TO-EXOPLANETS paradigm is motivated by the physical similarity of the atmospheres of these objects. Even if we think of a $0.020\ \mathcal{M}_\odot$ brown dwarf as a completely different class of object than Jupiter itself or a $5\ \mathcal{M}_{\mathrm{Jup}}$ exoplanet, we need many of the concepts developed by planetary scientists and applied by exoplanetary astronomers to model brown dwarf atmospheres. Similar temperatures and pressures exist in brown dwarf atmospheres and warmer gas giants, and therefore models of planetary atmosphere chemistry can be applied and tested in brown dwarfs. Condensation, clouds, vertical mixing, and winds have all been observed in brown dwarf atmospheres. The same techniques—even the same software codes—for modeling energy transport and climate are used for both brown dwarf and planetary atmospheres. How successful are these models, and can they be improved? Are we missing important physical processes in our models of exoplanets or brown dwarfs? In this paradigm, many use the term **worlds** to describe the planets, exoplanets, planetary-mass objects, and brown dwarfs that we are comparing. Brown dwarfs, as seen in Figure 1.3, have a significant advantage over exoplanets in that they are more luminous at a given age, and in turn their emission spectra can be more easily observed with standard stellar astronomy techniques. The new James Webb Space Telescope (JWST) is a particularly powerful tool for the study of atmospheres in worlds outside the Solar System, and we can confidently expect revolutionary advances in the next few years.

Finally, let us think about the Solar System planets, exoplanets, and brown dwarfs more generally. The planets of our Solar System are thought to have formed through a bottom-up, core accretion process, and collectively represent $<0.2\%$ of a solar mass. The observed statistics of exoplanets have shown that their numbers rise sharply to lower masses, and at least for "Hot Jupiters," their frequency depends strongly on the metallicity of their primordial disk. This formation scenario makes them distinctly different from our STAR-LIKE population. The desire to distinguish between brown dwarfs and planets is important enough for some astronomers that it is a key part of the International Astronomical Union Commission F2 "Exoplanets and the Solar System" working definition of an exoplanet (Lecavelier des Etangs & Lissauer 2022):

1. Objects with true masses below the limiting mass for thermonuclear fusion of deuterium (currently calculated to be 13 Jupiter masses for objects of solar metallicity) that orbit stars, brown dwarfs or stellar remnants and that have a mass ratio with the central object below the L_4 / L_5 instability:

$$\mathcal{M}/\mathcal{M}_{\text{central}} < 2/(25 + \sqrt{621}) \approx 1/25 \tag{1.2}$$

   are "planets," no matter how they formed. The minimum mass/size required for an extrasolar object to be considered a planet should be the same as that used in our Solar System.

2. Substellar objects with true masses above the limiting mass for thermonuclear fusion of deuterium are "brown dwarfs," no matter how they formed nor where they are located.

3. Free-floating objects in young star clusters with masses below the limiting mass for thermonuclear fusion of deuterium are not "planets," but are "sub-brown dwarfs" (or whatever name is most appropriate).

This working definition puts two groups of what we will call brown dwarf "planetary mass objects" in the non-planet category, implicitly because they seemingly conceptually fit better under the STAR-LIKE paradigm—the isolated planetary mass objects (Rule 3), and the companions that appear to have formed as secondary binaries to more massive stars or brown dwarfs (Rule 1's the mass ratio criteria), thought it explicitly includes worlds that might be interpreted under the STAR-LIKE paradigm in Rule 1. We also see that this definition would allow us to call an Earth-mass planet orbiting the very-low-mass star VB 10 (~0.09 $\mathcal{M}_\odot$) or the brown dwarf Luhman 16A (~0.033 $\mathcal{M}_\odot$), a planet rather than a moon or satellite. Luhman 16B, with nearly the same mass as Luhman 16A, is clearly a brown dwarf, but we would not call it a planet under this definition even if it were ~0.002 $\mathcal{M}_\odot$. The IAU commission would not call WISE J085510.83-071442.5 ($\mathcal{M} \lesssim 0.010$ $\mathcal{M}_\odot$)

itself a planet, and what we would call an Earth-mass object orbiting it is undefined. Chauvin et al. (2004)'s discovery of 2M1207b (or TWA 27B), often described as the first direct image of a gas giant planet outside the Solar System, does not qualify as a planet under Rule 1, since it is approximately 1/4 the mass of the primary brown dwarf 2MASSW J1207334-393254A (TWA 27A), and indeed it is generally considered to be too massive to have formed by core accretion like Jupiter. These proposed IAU definitions and concepts should be considered seriously, but whether or not they are useful in posing and solving research questions is a question for you, and widely disparate views are held by brown dwarf researchers. Pinfield et al. (2013) reports the results of a poll of attendees of the *Brown Dwarfs Come of Age* conference: some 67% considered 2M1207b a planet (violating Rule 1's mass ratio), 66% considered an object (κ And b) just above the deuterium-burning limit but with a mass of just 0.5% of its primary a planet (meeting Rule 1's mass ratio but violating Rule 2), 40% considered various free-floating planetary mass objects "planets" (violating Rule 3), and a small minority of 11% would even count a 0.02–0.05 $\mathcal{M}_\odot$ object, well above the deuterium-burning limit, which orbits a 0.60 $\mathcal{M}_\odot$ star as a planet (violating Rule 2). The takeaway point is that the term planet is legitimately a subject for discussion and motivates ongoing research.[2]

Ultracool dwarfs include all M7–M9 dwarfs, L dwarfs, T dwarfs, and Y dwarfs whether they are very-low-mass stars, brown dwarfs, planetary mass objects, or directly imaged "planets." Ultracool, of course, only applies from stellar astrophysics perspective, as it corresponds to $T_{\text{eff}} \lesssim 2700$ K. While we can hold different views on the use of terms like "planet," spectral types have specific definitions and meaning. They are based on actual discoveries, and not theoretical predictions.

1.3 Discovery of Brown Dwarfs

Today, thousands of very-low-mass stars and brown dwarfs have been observed with spectroscopy, with tens of thousands more detected in deep imaging surveys. To describe them, astronomers have extended Annie Jump Cannon's system of OBAFGKM stars to M7 dwarfs like VB8, down to L dwarfs, T dwarfs, and finally Y dwarfs, with the coldest known object, WISE J085510.83-071442.5, tentatively called Y4. However, the first observational discoveries that were widely accepted as brown dwarfs occurred in 1994–1995, three decades after the pioneering theoretical predictions. Why did it take so long? At the time of the theoretical discovery, the lowest luminosity stars known were VB 8 and VB 10, now called M7 and M8. Kumar (1963c) even proposed VB 10 may be a brown dwarf. Examining the theoretical calculations in Figure 1.2 and the empirical observed spectral type-luminosity relation shown, we can see the problems that faced astronomers. M7 or M8 dwarfs like VB8 or VB10 can be brown dwarfs if they are young enough, but they are more likely to be old hydrogen-burning stars. Without an independent age estimate, we cannot convincingly claim that they are brown dwarfs, and in the case

---

[2] A poll at a primarily exoplanet conference would surely find lower percentages considering any of these objects "planets."

of VB10 itself, the evidence favors an age $> 10^9$ Gyr. Proper motion surveys based on red-sensitive photographic plates identified more late-M dwarfs, and even objects like LHS 2924 that defined a new M9 type, but these discoveries lacked age information. This same problem also plagued interpretation of the pioneering 1988 discovery GD 165B by Becklin & Zuckerman (1988), an object that orbits a white dwarf and is much fainter and redder than M9 dwarfs—today we'd call it the first of the new L dwarf spectral class and classify it as L4. We can see in Figure 1.2 that if it is older than $\sim$2Gyr, it is predicted to be a $\sim$0.074 $\mathcal{M}_\odot$ very-low-mass star in this set of models.

Astronomers needed to either find a way to distinguish M-type brown dwarfs from M-type very-low-mass stars or to find an object so cool and faint that it could only be a brown dwarf. The first option was opened by the Rebolo et al. (1992)'s invention of the "lithium test:" Fully convective stars will burn all their lithium by the time they reach the main sequence. However, brown dwarfs below $\sim$0.06 $\mathcal{M}_\odot$ will never burn their lithium, so the presence of an atomic lithium absorption line at 670.8 nm in a cool M dwarf that is, say, at least a hundred million years old shows it is a brown dwarf. This line of research led to the confirmation that probably the M6.5 dwarf PPl 15 and certainly the M9 dwarf Teide 1 (Rebolo et al. 1995), both members of the Pleiades star cluster, were brown dwarfs. Around the same time, a faint companion to the nearby star Gliese 229 was discovered by Nakajima et al. (1995). Oppenheimer et al. (1995) found Gliese 229B's infrared spectrum has methane, proving it is 1000K or less (Tsuji 1964) and unambiguously a brown dwarf. It was later classified as a peculiar T7, and we can see in Figure 1.2 that it is between 0.020 $\mathcal{M}_\odot$ and 0.050 $\mathcal{M}_\odot$ even if it is older than 1 Gyr.

You might imagine following the STAR-LIKE paradigm to use radial velocity searches to find brown dwarf companions by their mass alone—after all, a large fraction of G dwarfs and M dwarfs have lower-mass stellar companions with orbital periods of days, months, or years. Although the sensitivity needed to detect brown dwarf companions in short orbits existed by the late 1980s, it turned out that such companions are surprisingly rare. The results were so bleak that this lack of companions became known as the "brown dwarf desert." A second difficulty is that this method is sensitive to $\mathcal{M} \sin i$ rather than the mass itself. For example, Latham et al. (1989) described their discovery of HD 114 762b with $\mathcal{M} \sin i = 0.011$ $\mathcal{M}_\odot$ as "probably a brown dwarf, and [it] may even be a giant planet"—but because there was evidence $\sin i$ is low for this star it was (and still is) generally thought to be face-on stellar companion with $\mathcal{M} > 0.08$ $\mathcal{M}_\odot$. Radial velocity searches eventually leapt past brown dwarfs into the planetary regime by the discovery of 51 Peg b by Mayor & Queloz (1995) with $\mathcal{M} \sin i = 0.000 45$ $\mathcal{M}_\odot = 0.47$ $\mathcal{M}_{\mathrm{Jup}}$ in an orbital period of 4.23 days. By a curious coincidence, Teide 1, Gl 229B, and 51 Peg b were all publicly announced at a special session of the *Cool Stars 9* conference in October 1995. From this date, the observational discoveries exploded. The numbers of exoplanets became hundreds and then thousands thanks to radial velocities surveys and later transit surveys. The numbers of brown dwarfs also became hundreds and then thousands as sky surveys like 1996–2001 Deep Near-Infrared Survey of the of the Southern Sky (DENIS, Epchtein et al. 1997), the 1997–2001 Two Micron All-Sky Survey (2MASS;

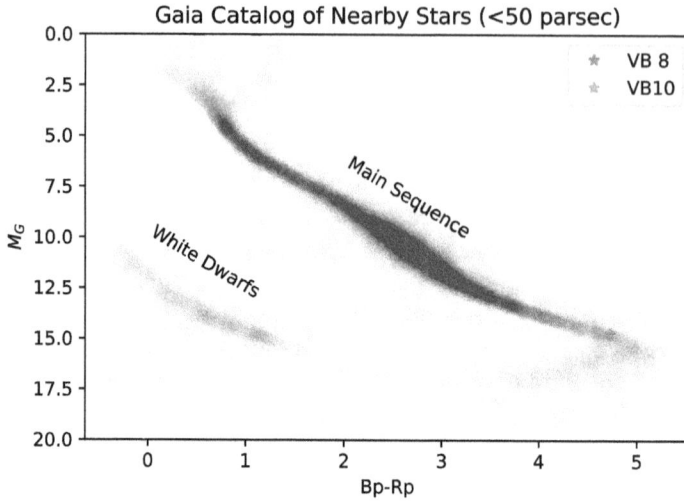

Figure 1.5. A density (Hess) version of the HR Diagram for the 50 parsec sample drawn from the Gaia Catalog of Nearby Stars (Gaia Collaboration et al. 2021). The prototype M7 and M8 dwarfs, VB8 and WB10, are shown as stars. The bluewards turn past VB10 is an observational artifact.

Skrutskie et al. 2006), 2000–2005 Sloan Digital Sky Survey (SDSS-I, Adelman-McCarthy et al. 2007), 2005–2012 UKIRT Infrared Deep Sky Survey (UKIDSS, Lawrence et al. 2007), and 2010 Wide-field Infrared Survey Explorer (WISE, Wright et al. 2010) enabled the discovery of nearby brown dwarfs by their photospheric emission. Dedicated searches in star-forming regions and clusters have discovered hundreds more brown dwarfs.

The advance of brown dwarf observations can best be seen in the observational HR diagrams. Figure 1.5 is based entirely on the optical photometry and parallax data reported for all stars and brown dwarfs within 50 pc in the Gaia Catalogue of Nearby Stars (Gaia Collaboration et al. 2021). Figure 1.6 is based on a compilation of near-infrared photometry and parallaxes of ultracool dwarfs, the UltracoolSheet (Best et al. 2020). In both, VB8 and VB10 are marked specially. The main sequence, which is a mass sequence of stars burning hydrogen, overlaps with and is continued by a brown dwarf cooling sequence. While the theoretical HR diagram (Figure 1.2) extends smoothly to cooler temperatures and lower temperatures—that is, always to the lower right, the two observational diagrams show much richer structure and at times bend to the left. For the near-infrared data, this has an important astrophysical cause—methane absorption makes the brown dwarfs look bluer in this diagram, even though they are cooler. For the Gaia data, this is an instrumental artifact—the brown dwarfs become so faint that they are not meaningfully detected in B, so the $B - R$ colors are really just lower limits. Indeed, most cooler L, T, and Y dwarfs within 50 pc are too faint to be measured by Gaia et al.[3] In the next chapter, we begin with an overview of the basic concepts of stellar astrophysics as applied to brown dwarfs.

---

[3] Here is an important lesson—public data releases are great, but always read the documentation!

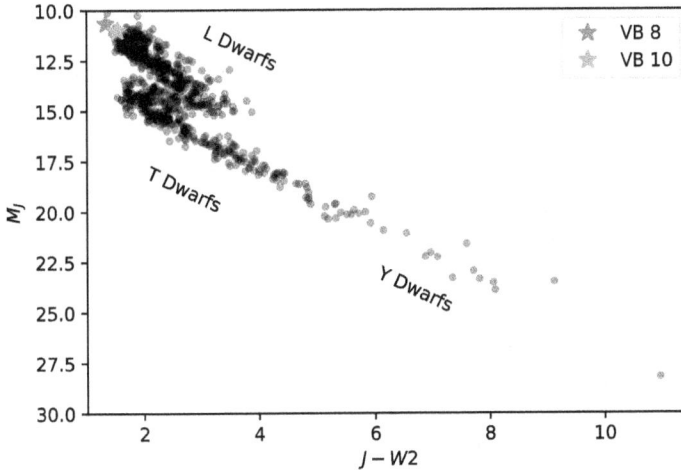

Figure 1.6. The observed HR Diagram for the brown dwarfs using ground-based J and space-based W2 photometry. Data compiled by Best et al. (2021). The prototype M7 and M8 dwarfs, VB8 and WB10, are shown as stars. The bluewards turn near $M_J \approx 14$ is a real astrophysical effect.

Ultracool Dwarf Data Compilation The UltracoolSheet is a compilation of photometry, astrometry, and spectral types for nearby very-low-mass stars, brown dwarfs, and directly imaged exoplanets. It is available at http://bit.ly/UltracoolSheet as a Google spreadsheet. It is maintained by Will Best, Trent Dupuy, Michael Liu, Rob Siverd, and Zhoujian Zhang and developed from compilations by Dupuy & Liu (2012), Dupuy & Kraus (2013), Deacon et al. (2014), Liu et al. (2016), Best et al. (2018, 2021), and Schneider et al. (2023)

Collaborative Brown Dwarf Database SIMPLE is a collaborative database for very-low-mass stars, brown dwarfs, and directly imaged exoplanets. It includes photometry, spectra, and modeled parameters. All data can be accessed through a web interface at https://simple-bd-archive.org but the full source code and archive can be installed on your computer using the instructions at https://github.com/SIMPLE-AstroDB/ SIMPLE-db/tree/main

---

Astronomical Database for Stars The SIMBAD Astronomical Database includes names, basic data, cross-identifications, and complete bibliographical references for nine million notable stars, including many nearby brown dwarfs. It is not a catalog and does not include every star that has been observed. It is an important complement for the ultracool dwarf UltracoolSheet and the Simple Database. Its name service is particularly useful for sorting through the often bewildering variety of catalog names for published ultracool dwarfs. https://simbad.u-strasbg.fr/simbad/

References

Adelman-McCarthy, J. K., Agüeros, M. A., Allam, S. S., et al. 2007, ApJS, 172, 634

Becklin, E. E., & Zuckerman, B. 1988, Natur, 336, 656

Best, W. M. J., Dupuy, T. J., Liu, M. C., Siverd, R. J., & Zhang, Z. 2020, The UltracoolSheet: Photometry, Astrometry, Spectroscopy, and Multiplicity for 3000+ Ultracool Dwarfs and Imaged Exoplanets, v1.0.0, Zenodo

Best, W. M. J., Liu, M. C., Magnier, E. A., & Dupuy, T. J. 2021, AJ, 161, 42

Best, W. M. J., Magnier, E. A., Liu, M. C., et al. 2018, ApJS, 234, 1

Chauvin, G., Lagrange, A., Dumas, C., et al. 2004, Astronomy and Astrophysics, 425, L29

Deacon, N. R., Liu, M. C., Magnier, E. A., et al. 2014, ApJ, 792, 119

Dupuy, T. J., & Kraus, A. L. 2013, Sci., 341, 1492

Dupuy, T. J., & Liu, M. C. 2012, ApJS, 201, 19

Eddington, A. S. 1926, The Internal Constitution of the Stars (Cambridge: Cambridge Univ. Press)

Epchtein, N., de Batz, B., Capoani, L., et al. 1997, Msngr, 87, 27

Esplin, T. L., & Luhman, K. L. 2019, AJ, 158, 54

Gaia CollaborationSmart, F. L., Sarro, L. M., et al. 2021, Astronomy and Astrophysics, 649, A6

Harris, C. R., Millman, K., van der Walt, S. J., et al. 2020, Natur, 585, 357

Hayashi, C., & Nakano, T. 1963, PThPh, 30, 460

Hunter, J. D. 2007, CSE, 9, 90

Kumar, S. S. 1962, AJ, 67, 579

Kumar, S. S. 1963a, ApJ, 137, 1121

Kumar, S. S. 1963b, ApJ, 137, 1126

Kumar, S. S. 1963c, AJ, 68, 283

Kumar, S. S. 2003, Proc. IAU Symp., (Vol. 211, ed. E. Martín; San Francisco, CA: ASP) 3

Latham, D. W., Mazeh, T., Stefanik, R. P., Mayor, M., & Burki, G. 1989, Natur, 339, 38

Lawrence, A., Warren, S. J., Almaini, O., et al. 2007, MNRAS, 379, 1599

Lecavelier des Etangs, A., & Lissauer, J. J. 2022, NewAR, 94, 101641

Liu, M. C., Dupuy, T. J., & Allers, K. N. 2016, ApJ, 833, 96

Low, C., & Lynden-Bell, D. 1976, MNRAS, 176, 367

Matthews, E. C., Carter, A. L., Pathak, P., et al. 2024, Natur, 633, 789

Mayor, M., & Queloz, D. 1995, Natur, 378, 355

Nakajima, T., Oppenheimer, B. R., Kulkarni, S. R., et al. 1995, Natur, 378, 463

Nakano, T. 2014, 50 Years of Brown Dwarfs (Vol 401, Switzerland: Springer International) 5

Oppenheimer, B. R., Kulkarni, S. R., Matthews, K., & Nakajima, T. 1995, Sci., 270, 1478

Payne, C. H. 1925, PhD thesis, Radcliffe College

Pinfield, D. J., Beaulieu, J. P., Burgasser, A. J., et al. 2013, MmSAI, 84, 1154

Rebolo, R., Martin, E. L., & Magazzu, A. 1992, ApJL, 389, L83

Rebolo, R., Zapatero Osorio, M. R., Martín, E. L., et al. 1995, Natur, 377, 129

Salpeter, E. E. 1955, ApJ, 121, 161

Sanghi, A., Liu, M. C., Best, W. M. J., et al. 2023, ApJ, 959, 63

Schneider, A. C., Munn, J. A., Vrba, F. J., et al. 2023, AJ, 166, 103

Silk, J. 1977, ApJ, 214, 152

Skrutskie, M. F., Cutri, R. M., Stiening, R., et al. 2006, AJ, 131, 1163

Tarter, J. 2014, 50 Years of Brown Dwarfs (Vol 401, Switzerland: Springer International) 19

Tarter, J. C. 1976, BAAS, 8, 517

Tsuji, T. 1964, PJAB, 40, 99

van Biesbroeck, G. 1944, AJ, 51, 61

van Biesbroeck, G. 1961, AJ, 66, 528

Virtanen, P., Gommers, R., Oliphant, T. E., et al. 2020, NatMe, 17, 261

Wright, E. L., Eisenhardt, P. R. M., Mainzer, A. K., et al. 2010, AJ, 140, 1868

An Introduction to Brown Dwarfs

From very-low-mass stars to super-Jupiters

John Gizis

Chapter 2

The Basics

2.1 Fundamental Parameters

We begin by considering the luminosity. For an isolated brown dwarf, the energy emitted is transported from the deep interior; unlike the Solar System planets, there is no external energy source in the form of sunlight. The luminosity is the total energy emitted—it has units of J s^{-1} (SI) or erg s^{-1} (cgs). It is often convenient to work in units of solar luminosity, which were defined by the IAU as a nominal solar luminosity 1 $\mathcal{L}_\odot = 3.828 \times 10^{26}$ J s^{-1} (Mamajek et al. 2015).

In principle, the luminosity of a brown dwarf can be determined if we measure its distance (d) and observe its spectral energy distribution at all wavelengths using the inverse square law. (We also assume, for now, spherical symmetry and that interstellar absorption is negligible):

$$F = \frac{\mathcal{L}}{4\pi d^2} \tag{2.1}$$

In terms of the flux density observed per unit frequency (F_ν, with SI units J s^{-1} m^{-2} Hz^{-1}) or per unit wavelength (F_λ, J s^{-1} m^{-2} μm^{-1})

$$\mathcal{L} = 4\pi d^2 \int_0^\infty F_\nu \, d\nu = 4\pi d^2 \int_0^\infty F_\lambda \, d\lambda \tag{2.2}$$

We are also interested in the flux at the "surface" of the brown dwarf ($\mathcal{F}$) with radius $\mathcal{R}$:

$$\mathcal{F} = \frac{\mathcal{L}}{4\pi \mathcal{R}^2} = F\left(\frac{d}{\mathcal{R}}\right)^2 \tag{2.3}$$

If brown dwarfs were perfect blackbodies with temperature T, then the surface flux density as a function of frequency ($\mathcal{F}_\nu$) or wavelength ($\mathcal{F}_\lambda$) could be calculated using the familiar Planck Function (B_ν or B_λ):

doi:10.1088/2514-3433/ad757ech2

$$\mathcal{F}_\nu = \pi B_\nu = \pi \; \frac{2h\nu^3}{c^2} \frac{1}{e^{h\nu/k_B T} - 1} \tag{2.4}$$

Integrating over all frequencies, and using the Stefan–Boltzmann constant (σ_{SB}) to simplify the expression

$$\mathcal{L} = 4\pi \mathcal{R}^2 \int_0^\infty \mathcal{F}_\nu d\nu = 4\pi \mathcal{R}^2 \sigma_{SB} T^4 \tag{2.5}$$

We can also work in terms of wavelength bins by using

$$B_\lambda = B_\nu \left| \frac{d\nu}{d\lambda} \right| = B_\nu \frac{c}{\lambda^2} \tag{2.6}$$

so that

$$\mathcal{F}_\lambda = \pi \frac{2hc^2}{\lambda^5} \frac{1}{e^{hc/\lambda k_B T} - 1} \tag{2.7}$$

However, **brown dwarfs are not blackbodies!** The Planck function is a very poor model for the spectral energy distribution. Nevertheless, we will define an effective temperature (T_{eff}):

$$T_{eff} \equiv \left(\frac{\mathcal{L}}{4\pi \mathcal{R}^2 \sigma_{SB}} \right)^{1/4} \tag{2.8}$$

The task of theorists is to predict the surface flux density ($\mathcal{F}_\lambda$) as a function of wavelength. To predict this spectral energy distribution, perhaps the most important parameter is the total surface flux ($\mathcal{F}$). This describes the power per unit area entering the bottom of the atmosphere, which is equal to the power per unit area of the energy radiated into space. Rather than report this value, which varies over many orders of magnitude from the oldest brown dwarfs to the Sun, we instead use the effective temperature. The total surface flux is then $\sigma_{SB} T_{eff}^4$; but while this effective temperature has units of K, we must always bear in mind that the actual temperature of the brown dwarf is a function of height in the atmosphere. This is the pressure–temperature (PT) profile, which we discuss in Chapter 6.

The task of observers is to observe the flux density (F_λ) as a function of wavelength. This is challenging for the Earth, especially for cooler brown dwarfs, because the Earth's atmosphere blocks some infrared wavelengths (Figure 2.1) and because the Earth is warm enough that the background becomes prohibitively bright. JWST however overcomes both these problems and features an array of instruments that can measure brown dwarfs from 1 to 20 μm. We look at the first two published JWST spectra.

Miles et al. (2023) presented the first published JWST brown dwarf spectrum. VHS J125601.58−125730.3b, or VHS1256−1257b, is a companion to a young M dwarf system. Discovered by Gauza et al. (2015) and at a projected separation of ~100 au, it is in effect isolated. We plot the observed JWST spectrum from 0.9 to 20 μm in Figure 2.2. Adopting the Gaia distance of the primary, $d = 21.15$ pc, and

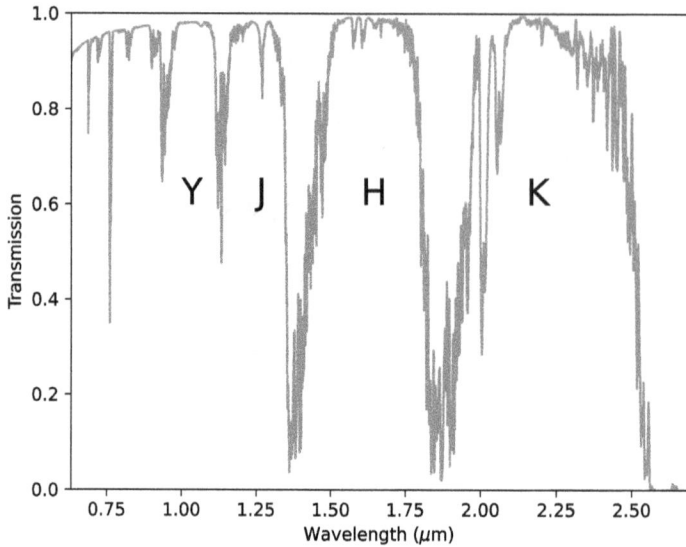

Figure 2.1. A representative example of transmission at a ground-based observatory (Cerro Paranal, European Southern Observatory). The major observable infrared regions (J,H,K) lie in between telluric water bands. The ESO sky and transmission model is described by Noll et al. (2012) and Moehler et al. (2014)

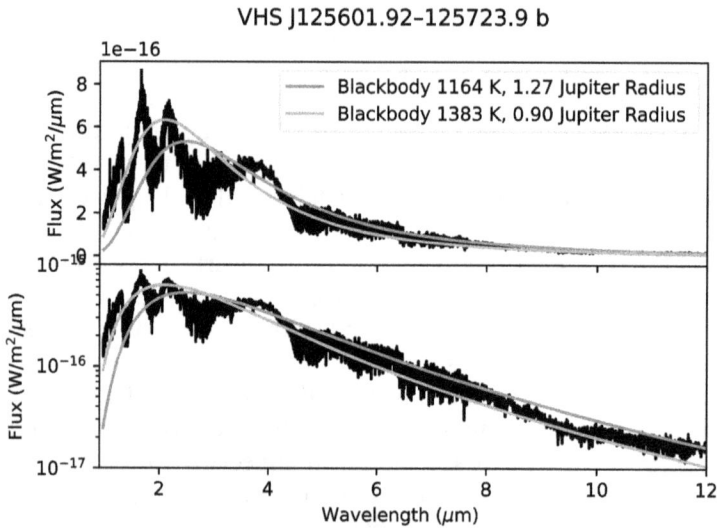

Figure 2.2. The observed JWST spectrum of the brown dwarf VHS1256b, a candidate planetary mass object. This is the first brown dwarf spectrum published using JWST. Also shown are blackbody spectra with the same luminosities for the preferred radius of 1.27 $\mathcal{R}_{\mathrm{Jup}}$ and, for comparison only, a smaller radius 1 $\mathcal{R}_{\mathrm{Jup}}$. The logarithmic scale can be deceiving: All three spectra have the same area ($=F$) under the curve.

numerically integrating the observed values of F_λ, we obtain $\mathcal{L} = 1.07 \times 10^{22}$ W. This should be increased by ~2% to account for the shorter and longer wavelength flux. Although the luminosity is strongly constrained, the effective temperature also

WISE J035934.06-540154.6

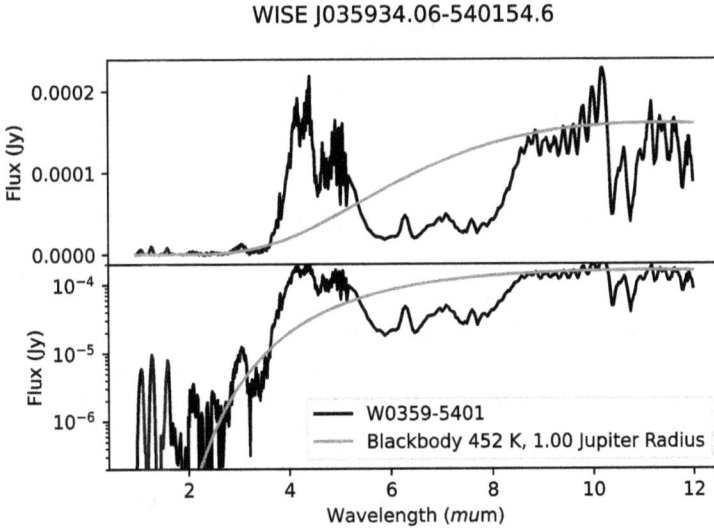

Figure 2.3. The observed JWST spectrum of the Y0 brown dwarf W0359-5401 (Beiler et al. 2023a). This is the second brown dwarf spectrum published using JWST. Also shown is a blackbody spectrum with the same luminosity for the representative radius $1 R_J$.

depends on the radius, which must be estimated using models. We show in Figure 2.2 a blackbody spectrum for an adopted radius of $1.27 \, \mathcal{R}_{\mathrm{Jup}}$, favored by the analysis of Miles et al., but also for comparison the spectrum for an adopted radius of $1.00 \, \mathcal{R}_{\mathrm{Jup}}$, typical of an older brown dwarf. We can see that neither blackbody looks very much like the observed spectrum.[1]

The Astropy Project Throughout this book, we will make use of the astropy Python package (Astropy Collaboration et al. 2013, 2018, 2022). It provides an essential ability to handle data tables, file input/output, blackbody models, conversion of units, physical constants, coordinate systems, and Solar System positions. It is available at https://www.astropy.org/index.html.

The first published JWST Y dwarf spectrum was presented by Beiler et al. (2023b). We plot this spectrum, which was published as F_ν, in Figure 2.3. It extends from 0.96 to 12.0 μm. JWST photometry, not shown, extended this spectral energy distribution out to 21 μm. We can numerically integrate the spectrum shown in Figure 2.3 to find $F = 5.5 \times 10^{17} \, \mathrm{J \, s^{-1} \, m^{-2}}$. However, the longer wavelength data is

---

[1] We can also see the fortunate fact that in the near-infrared, the brown dwarf flux peaks correspond to the J, H, K windows we can observe from Earth (Figure 2.1) because both atmospheres have H_2O molecules. Furthermore, the water absorption bands are broader in brown dwarfs than in the Earth, so we will see that they can be characterized by ground-based spectra.

not negligible. Including the photometric region out to 21 μm and extrapolating the longer wavelengths brings us to $F = 6.89 \times 10^{17}$ J s^{-1} m^{-2}. (The shorter wavelengths are negligible.) Adopting the distance of $d = 13.57$ pc, $\mathcal{L} = 1.52 \times 10^{20}$ W $= 3.93 \times 10^{-7} \mathcal{L}_\odot$. Adopting a radius of 1 $\mathcal{R}_{\text{Jup}}$, the blackbody spectrum for the effective temperature of 452 K is also shown in Figure 2.3.

2.2 Astronomy Units: Magnitudes and Parsecs

The SI or cgs system is not particularly well suited to astronomical scales. Brown dwarf work follows stellar astronomy in using magnitudes and custom units of distance. Apparent magnitudes (m) are used to describe the observed brightness (flux density) of stars and the absolute magnitude (M) is related to their intrinsic brightness (luminosity). Bolometric magnitudes represent the total flux (or "heat flux density") from the star. Unfortunately, various authors used different definitions of the solar luminosity or solar absolute bolometric magnitude, so it was not always clear what they mean by M_{bol} or m_{bol}. In 2015, the IAU therefore officially defined absolute bolometric magnitudes as

$$M_{\text{Bol}} \equiv -2.5 \log \left(\mathcal{L}/3.0128 \times 10^{28} \text{ W} \right) \tag{2.9}$$

This definition, along with the nominal solar luminosity, means that the Sun has $M_{\text{Bol} \odot} = 4.740$. An isotropic source with $M_{\text{Bol}} = 0$ at a 10 parsec has apparent magnitude $m_{\text{Bol}} = 0$, and in general

$$m = M + 5 \log \left(d/10 \text{ pc} \right) + A \tag{2.10}$$

The extinction (A) is usually negligible for brown dwarfs within 100 parsecs at typical wavelengths $\lambda > 800$ nm, but is very important even in the near-infrared when studying brown dwarfs in more distant star-forming regions. The astronomical unit (au), traditionally the average Earth–Sun distance, is now defined by the IAU as

$$1 \text{ au} \equiv 149\,597\,870\,700 \text{ m} \tag{2.11}$$

One parsec, the distance where the parallax angle (p) is one arcsec, is defined as 1 pc $\equiv 648\,000/\pi$ au so that 1pc $\approx 3.085\,678 \times 10^{16}$ m, and the distance in parsecs is $1/p$. This means that the apparent bolometric magnitudes are

$$m_{\text{Bol}} = -2.5 \log \left(F/2.518021 \times 10^{-8} \text{ W m}^{-2} \right) \tag{2.12}$$

We cannot measure all wavelengths simultaneously for brown dwarfs with a bolometer. Instead, astronomers measure magnitudes using images taken through different filter systems, a process known as photometry. We have already seen in Figures 1.5 and 1.6 absolute magnitudes and colors (magnitude differences) for the Gaia satellite's G, B_P, and R_P filters between 0.4 and 0.9 μm, the ground-based filter J filter around 1.2 μ m, and the WISE satellite's W2 filter around 4.6 μm. The main near-infrared magnitude bands (Y, J, H, K) were chosen to lie in relatively transparent windows between the major water absorption bands of Earth's atmosphere (Figure 2.1). The difference between the bolometric magnitude and the magnitude in a given filter is the bolometric correction (BC):

$$m_J + BC_J = m_{Bol} \tag{2.13}$$

A warning: Even though physically the luminosity through a filter is only a fraction of the bolometric luminosity, because of the way the magnitude zero-points are defined, the bolometric correction may be positive or negative! The position as a function of time, including the parallax and proper motion of brown dwarfs, can also be measured by these images, a process known as astrometry. The subtleties of photometry and astrometry are important enough that we discuss them in detail in Chapter 4.

2.3 Spectroscopy: Velocities, Surface Gravity and Metallicity

Much more information is encoded in the spectrum than just the effective temperature. The Y dwarf spectrum shown in Figure 2.3 has a resolution $R = \lambda/\Delta\lambda \approx 100$. Figure 2.4 shows much higher resolution ($R \approx 20,000$) spectra for two L dwarfs. The radial velocity ($v_{rad} = c(\lambda_{obs} - \lambda_{rest})/\lambda_{rest}$) can be measured from high-resolution spectra. Rotational broadening can also be measured; if the equatorial velocity is V and a star is inclined at angle i with $i = 0$ pole-on to Earth, then the parameter $V \sin i$ can be measured from the broadening of the lines. Care must be taken to model the effects of rotational broadening, pressure broadening,

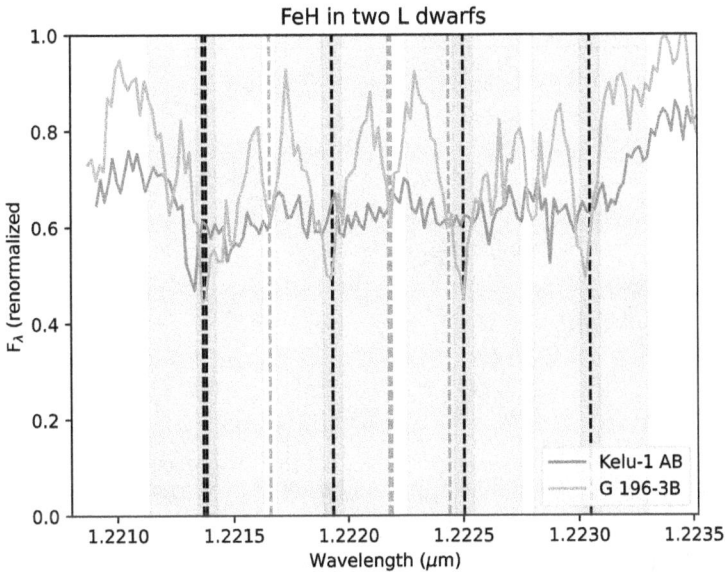

Figure 2.4. Brown Dwarf Spectroscopic Survey (BDSS) observations of the L dwarfs Kelu-1AB and G 196-3B (McLean et al. 2007). Spectra of this resolution can be used to measure the radial velocity (v_{rad}), which has been removed to show the spectra in the laboratory reference frame and projected rotational velocity ($V \sin i$). The slowly rotating ($V \sin i \approx 10$ km s^{-1}) G 196-3B shows sharp features due to the FeH transitions (vertical dashed lines). In the rapidly rotating Kelu-1 AB ($V \sin i \approx 60$ km s^{-1}), the individual lines are blended together into a broad feature. The vertical shaded regions mark ± 10 km s^{-1} and ± 60 km s^{-1} as guidance for some of the FeH lines. The instrumental resolution is $R \approx 20,000$ (15 km s^{-1}).

and instrumental resolution. Figure 2.4 shows two brown dwarfs with similar temperature: Kelu-1's spectrum is broadened by rapid rotation while G 196-3B has sharp features at the instrumental resolution. For a given rotation period (P), the equatorial velocity is

$$V = \frac{2\pi \mathcal{R}}{P} = 30.4 \text{ km s}^{-1} \left(\frac{\mathcal{R}}{0.1\,\mathcal{R}_\odot}\right)\left(\frac{4\text{ h}}{P}\right) \tag{2.14}$$

The Morgan–Keenan (MK) spectral classification system (Morgan & Keenan 1973) uses the appearance of atomic lines to distinguish between dwarfs (V), giants (III), and supergiants (I). This works through the link between the pressure in the photosphere and the surface gravity, but remember that these classes have radii that vary by orders of magnitude, much more than the variation among ultracool dwarfs. The surface acceleration due to gravity, g, is

$$g = \frac{G\mathcal{M}}{\mathcal{R}^2} \tag{2.15}$$

In the cgs system (cm s^{-2}), the Sun has $\log g = 4.4$ and Jupiter has $\log g = 3.4$. Figure 2.5 shows the predicted effective temperature and surface gravity for the MESA models from Chapter 5. Most ultracool dwarfs ($0.03\,\mathcal{M}_\odot \leqslant \mathcal{M} \leqslant 0.09\,\mathcal{M}_\odot$)

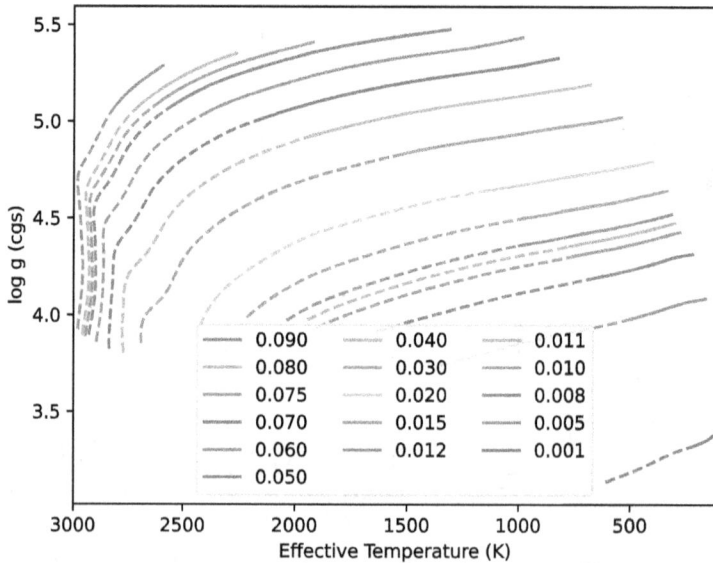

Figure 2.5. Model predictions for T_{eff} and $\log g$ for the evolutionary models calculated in Chapter 5. The stars and brown dwarfs evolve from high temperature to low temperature (left to right) and low gravity to higher gravity (upwards), and more massive model tracks are always above and to the left of lower mass tracks. The model evolutionary tracks are shown as dashed lines for the first 100 Myr and as solid lines from 100 Myr to 10 Gyr.

older than 100 Myr have $\log g \approx 5.0 \pm 0.2$.[2] A precise measurement of g offers the potential to distinguish lower mass objects from higher mass ones, and young objects with large radius from fully contracted ones. Measuring pressure broadening in ultracool dwarfs with high resolution is one possible strategy to measure g. In Figure 2.4; G 196-3B is a young L dwarf expected to have lower g than most L dwarfs. We observe sharp lines at the limit of the instrumental resolution but we are aided by the apparent low projected rotational velocity: In many L dwarfs, the rotational broadening dominates over any pressure/gravity effect. However, surface gravity also affects the near-infrared spectral energy distribution is ways that are apparent even at low resolution ($R \approx 100$), and as we will see in Chapter 3, differences of ~ 0.5 in $\log g$ can be identified. If we can reach precisions of ± 0.1 or better, we would have a powerful tool for constraining brown dwarf masses.

The spectra shown in Figures 2.2 and 2.3 show absorption by molecules CO, CH_4, H_2O, and many others. If we are to understand the evolution or the chemistry of brown dwarf atmospheres, we need to know the relative number of each element H, C, and O. A good starting point in the STAR-LIKE paradigm is the solar abundance, where astronomers have both spectroscopy and direct analysis of meteorite samples to guide them. The number of atoms of each element is reported relative to 10^{12} atoms of hydrogen. For example:

$$A(\text{Fe}) = \log N(\text{Fe})/N(\text{H}) + 12 \qquad (2.16)$$

An observed value $A(\text{Fe}) = 7.50$ means that there are 32,000,000 iron atoms for every trillion (10^{12}) hydrogen atoms in the solar photosphere. Table 2.1 gives abundances for the solar photosphere as reported by Anders & Grevesse (1989) (AG89), Asplund et al. (2009) (A09), Lodders (2010) (L10), and (Magg et al. 2022)

Table 2.1. Solar Photosphere Abundances for Selected Elements

Element	A (AG89)	A (A09)	A (L10)	A (M22)
H	12	12	12	12
C	8.56 ± 0.04	8.43 ± 0.05.	8.39 ± 0.04	8.56 ± 0.05
N	8.05 ± 0.04	7.83 ± 0.05.	7.86 ± 0.12	7.98 ± 0.08
O	8.93 ± 0.035	8.69 ± 0.05	8.73 ± 0.07	8.77 ± 0.04
Mg	7.58 ± 0.05	7.60 ± 0.04	7.54 ± 0.06	7.55 ± 0.05
Si	7.55 ± 0.05	7.51 ± 0.03	7.53 ± 0.06	7.59 ± 0.07
K	5.12 ± 0.13	5.03 ± 0.09	5.11 ± 0.04	5.14 ± 0.10
Ti	4.99 ± 0.02	4.95 ± 0.05	4.93 ± 0.03	4.94 ± 0.05
V	4.00 ± 0.02	3.93 ± 0.08	3.99 ± 0.03	3.89 ± 0.08
Fe	7.67 ± 0.03	7.50 ± 0.04	7.46 ± 0.08	7.50 ± 0.06

---

[2] We will follow the older astronomical literature that assumes cgs, so that if we refer to $\log g = 5$ without specifying units, we mean $g = 10^5$ cm s^{-2} = 10^3 m s^{-2}.

(M22) for a few elements that we will observe in brown dwarfs. The AG89 values are obsolete but representative of how the older literature used a higher solar abundance than more recent work. The values listed are for the present-day solar photosphere: A for all elements heavier than He should be increased by 0.053 (13%) to represent the initial composition of the Sun before gravitational settling, but this does not change the relative abundance of the heavy elements (Lodders 2010). For models of the interiors of stars and brown dwarfs, we are interested in the mass fraction of hydrogen (X), helium (Y), and all other elements (Z), called "metals" in the astronomy tradition. The ratio Z/X can be obtained by summing all the observed values of A and weighting by their mass. The present-day photospheric values of X, Y, and Z are then determined by this ratio and agreement with solar seismology models that are sensitive to Y. We have to note that the generally accepted overall metallicity of the Sun (Z) has varied over the last four decades, and in particular the carbon, nitrogen, and oxygen abundances in the Sun have been controversial. These elements have few observable photospheric lines, so complex three-dimensional solar atmosphere models are needed to interpret the observations, and meteorites do not preserve the original abundance of C, N, or O. Table 2.1 are logarithmic so that the number of carbon atoms relative to hydrogen varies by 50%! For brown dwarfs, the carbon-to-oxygen ratio is a critical parameter for the chemistry: The A09 value is $0.55^{+0.10}_{-0.08}$ but the often-used L10 value is 0.46 and the newer M22 value is 0.62. The M22 values give $Y = 0.2734$ and $Z = 0.0176$ for the initial solar composition, which coincidentally is close to the generally accepted values used in the 1990s, but the often-used A09 value is much lower at $Z = 0.0154$. The lesson here is that the meaning of "solar composition" may be different in various studies, and if the most recent studies are correct, there may once again be significant updates to C, N, and O.

Most stars have abundances that are similar to the solar composition, and indeed, abundances in other stars are measured relative to the solar composition, so that:

$$[\text{Fe/H}] = \log(N(\text{Fe})/N(\text{H}) - \log(N(\text{Fe})/N(\text{H}))_\odot \qquad (2.17)$$

Figure 2.6 shows the measured [Fe/H] and [O/Fe] = [O/H] − [Fe/H] for 1018 G dwarfs and subgiants within 50 parsecs of the Sun compiled in the Hypatia Catalog (Hinkel et al. 2014). These measurements are based on spectra with $30,000 < R < 120,000$. This is only a representative plot to illustrate the fact that most nearby stars are close to (but not exactly!) scaled solar abundances, with overall metallicity for most nearby stars within [Fe/H] $\approx 0 \pm 0.3$—"only" a factor of 2! For the critical C/O ratio, Brewer & Fischer (2016) concluded that the median C/O ratio in nearby FGK dwarfs is 0.47, less than their adopted (A09) solar value of 0.55. Remarkably, this suggests that use of the lower Lodders (2010) C/O ratio is well justified for a sample of local brown dwarfs. There is a trend of C/O with [Fe/H] so that metal-poor disk stars have lower C/O. They also report that none of their 849 stars have C/O >0.7, which for our SUPER-JUPITER paradigm suggests that a reliably determined high C/O in an ultracool dwarf may be sign of a planetary formation scenario. As seen in the figure, metal-poor stars with [Fe/H] $\lesssim 0.5$

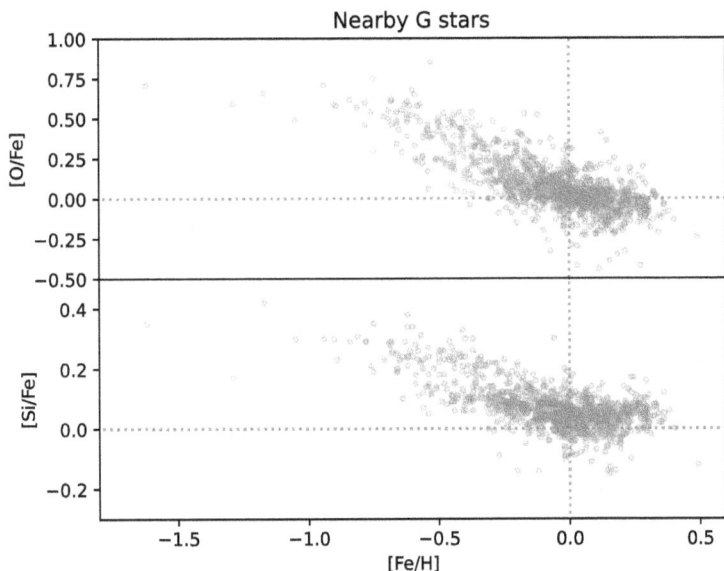

Figure 2.6. Abundances for iron, oxygen, and silicon in nearby G dwarfs compiled by the Hypatia Catalog (Hinkel et al. 2014).

typically have very non-solar compositions because the alpha-capture elements (O, Ne, Mg, Si, S, Ar, Ca, Ti) are primarily produced in massive-star supernovae but the iron-peak elements are primarily produced later by white-dwarf supernovae, so that $[\alpha/\mathrm{Fe}] > 0.2$. About 2% of stars within 50 pc locally are part of the old and mildly metal-poor Galactic thick disk and about 0.2% are the true old, metal-poor Population II Galactic halo. For our purposes, it is worth considering that the APOGEE survey (Anguiano et al. 2020) chemically defined the thick disk as metal-poor but α enhanced ($-1 < [\mathrm{Fe/H}] < 0$, $+0.18 < [\mathrm{Mg/Fe}] < +0.4$ and the halo as $[\mathrm{Fe/H}] < -1$, although Gaia has shown complex chemical and kinematic substructures for the old disk, thick disk, and halo far beyond the level that we will be able to see in local samples of brown dwarfs for the near future.

The techniques for measuring T_{eff}, $\log g$, and abundances for FGK dwarfs from high-resolution spectra are well established even if it should be clear that uncertainties remain. How to measure T_{eff}, $\log g$, $[\mathrm{Fe/H}]$, C/O, $[\mathrm{O/H}]$, and other abundances for nearby brown dwarfs is a key problem for the rest of this book! In the future, we might imagine that analysis of ultracool dwarfs will become so routine and reliable that astronomers will proceed directly from the observed spectra to these parameters: A goal for our field might be to match a catalog like the APOGEE Data Release 17 (Majewski et al. 2017; Abdurro'uf et al. 2022) that includes near-infrared spectra of more than 300,000 red giant stars and reports measurements of T_{eff}, $\log g$, and overall metallicity $[M/\mathrm{H}]$ and many specific elemental abundances such as $[\mathrm{Mg/H}]$. We are not yet at that stage, and so the next chapter describes how and why we use the spectral classification system of M, L, T, and Y dwarfs.

References

Abdurro'uf, E., Accetta, K., Aerts, C., et al. 2022, ApJS, 259, 35

Anders, E., & Grevesse, N. 1989, GeCoA, 53, 197

Anguiano, B., Majewski, S. R., Hayes, C. R., et al. 2020, AJ, 160, 43

Asplund, M., Grevesse, N., Sauval, A. J., & Scott, P. 2009, ARA&A, 47, 481

Astropy Collaboration, Robitaille, T. P., Tollerud, E. J., et al. 2013, A&A, 558, A33

Astropy Collaboration, Price-Whelan, A. M., Sipőcz, B. M., et al. 2018, AJ, 156, 123

Astropy Collaboration, Price-Whelan, A. M., Lim, P. L., et al. 2022, ApJ, 935, 167

Beiler, S., Cushing, M., Kirkpatrick, D., et al. 2023a, ApJL, 951, L48

Beiler, S. A., Cushing, M. C., Kirkpatrick, J. D., et al. 2023b, ApJL, 951, L48

Brewer, J. M., & Fischer, D. A. 2016, ApJ, 831, 20

Gauza, B., Béjar, V. J. S., Pérez-Garrido, A., et al. 2015, ApJ, 804, 96

Hinkel, N. R., Timmes, F. X., Young, P. A., Pagano, M. D., & Turnbull, M. C. 2014, AJ, 148, 54

Lodders, K. 2010, Principles and Perspectives in Cosmochemistry (Berlin: Springer-Verlag) 379

Magg, E., Bergemann, M., Serenelli, A., et al. 2022, A&A, 661, A140

Majewski, S. R., Schiavon, R. P., Frinchaboy, P. M., et al. 2017, AJ, 154, 94

Mamajek, E. E., Prsa, A., Torres, G., et al. 2015, arXiv:1510.07674

McLean, I. S., Prato, L., McGovern, M. R., et al. 2007, ApJ, 658, 1217

Miles, B. E., Biller, B. A., Patapis, P., et al. 2023, ApJL, 946, L6

Moehler, S., Modigliani, A., Freudling, W., et al. 2014, A&A, 568, A9

Morgan, W. W., & Keenan, P. C. 1973, ARA&A, 11, 29

Noll, S., Kausch, W., Barden, M., et al. 2012, A&A, 543, A92

Chapter 3

Spectral Types

3.1 Spectral Typing Strategy

The purpose of a spectral classification system is to provide us with an empirical language to discuss and compare observations of brown dwarfs. It is empirical because observed spectra—not theoretical synthetic spectra, nor theoretical interpretations of observed spectra— have used to define the spectral types. Even if we had an apparently reliable theory, it is worth bearing in mind the advice of stellar atmosphere theorist Dimitri Mihalas when asked to comment on the Morgan–Keenan (MK) spectral classification system for OBAFKM stars (Morgan & Keenan 1973):

> One of the most insidious problems that can arise is when we deal with a theoretical system that is seemingly self-consistent *and yet is incorrect*. Here, we may, for example, make diagnoses of (say) the temperature and gravity of a star which follows with a small error (of comparison) from the theory, and *yet are systematically wrong!* In the final analysis, there is only one meaningful approach that we can adopt in the empirical system: to *define it in terms of real objects, without comment.*. For example, the most valid way to "describe" a stellar spectrum is to *refer* it to spectra of prechosen *standards* ....

The use of spectral classification is not a weakness in our understanding, but a powerful tool for analysis! The vast majority of the thousands of ultracool dwarf spectra that been observed can be simplified into a smaller number of categories. The main ultracool dwarf classification systems have been inspired by the MK process, so that they identify specific objects as primary spectral standards in order to define each subtype for what are now known as M, L, T, and Y dwarfs. L dwarf classification systems were first developed using the "optical" meaning 630 nm $< \lambda <$ 1000 nm by Martín et al. (1999) and Kirkpatrick et al. (1999). In this chapter, we will mainly show $R \approx 900$ (Keck) and $R \approx 2000$ (Sloan Digital

Table 3.1. Spectral Standard References

Types	Wavelengths	Region	Reference
M0–M9	620–900 nm	"optical"	Kirkpatrick et al. (1991)
L0–L8	650–1000 nm	"optical"	Kirkpatrick et al. (1999)
T0–T8	0.8–2.5 μm	"infrared"	Burgasser et al. (2006)
T9, Y0, Y1	0.8–2.5 μm	"infrared"	Cushing et al. (2011)
Y1	0.8–2.5 μm	"infrared"	Kirkpatrick et al. (2012)
T8-Y1	1–12 μm	"infrared"	Beiler et al. (2024)

Sky Survey) optical spectra. For T and Y dwarfs, the primary classification system is defined in the "near-infrared," or $1.0~\mu$m $< \lambda < 2.4~\mu$m. M and L dwarfs can also be classified in the near-infrared. We will primarily show near-infrared spectra from the SpeX Prism with $R \approx 75 - -200$.

The $M/L/T/Y$ system is a one-parameter (one-dimensional) sequence, and we expect that the main underlying physical parameter is T_{eff}. Ultimately, classification is based on all the atomic and molecular features in the chosen wavelength range, not just one or two special ones, and this enables "peculiar" objects to also be identified for further study. References that include the primary spectral standards are listed in Table 3.1. One key difference from the original MK process is that the MK classes are a two-dimensional system with a luminosity class that uses surface-gravity (g) sensitive features to distinguish between main-sequence dwarf stars (Class V) and evolved subgiants (IV), giants (III), and super-giants (II, Ia, Ib). These evolutionary stages do not exist for brown dwarfs: all of our objects are simply considered dwarfs (class V). However, the STAR-LIKE paradigm expects many objects with either low surface gravity (Figure 2.5) or low metallicity (Figure 2.6). Moving from listing these objects as peculiar outliers to developing full-fledged systems to characterize young, low-gravity objects and metal-poor "subdwarfs" has been an important development for the field and is still ongoing.

3.2 M and L dwarfs

Kirkpatrick et al. (1991) presented and extended the MK system to the red ($630 - 900$ nm) and defined standards for M0V to M9V dwarfs: M6's standard is Wolf 359; M7 is VB8; M8 is VB10; M9 is LHS 2924. Later, Kirkpatrick et al. (1999) defined standards over 650–1000 nm for L0V to L8V. These systems based on the appearance of the spectra at $<1~\mu$m will be called "optical" classification. Let us consider a practical example of what this means. Figure 3.1 shows observed spectra of five ultracool dwarfs—Kelu-1, 2MASSW J0030438+313932, 2MASSW J0753321+291711, 2MASSW J0829066+145622, and 2MASSW J0928397-160312 — taken with the same observing setup (Kirkpatrick et al. 1999, 2000). These are rich spectra with numerous molecular absorption features. Some small differences between these five objects are evident, but it should be clear that all five objects have very similar spectra. Kirkpatrick et al. (1999) defined Kelu-1 as the primary standard

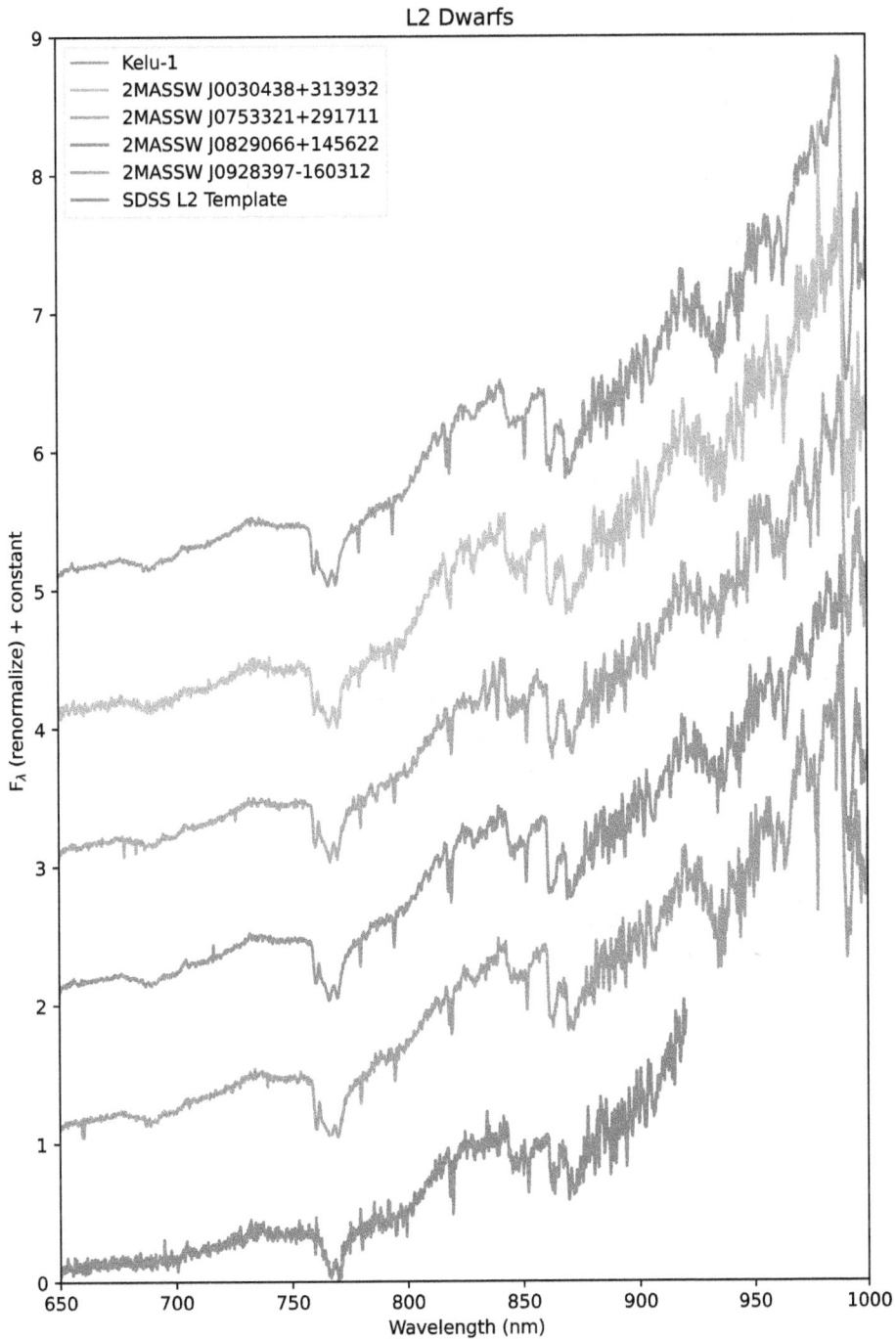

Figure 3.1. Five objects classified by L2 by Kirkpatrick et al. (1999, 2000). Also shown is a template L2 spectrum created by Schmidt et al. (2014) using the mean of seven different L2 dwarfs observed by the Sloan Digital Sky Survey (SDSS).

that defines type "L2V," so that all these stars have been classified as L2 dwarfs. Let us emphasize that this is a purely observational language: "2MASSW J0753321 +291711 is classified as a L2 dwarf in the optical" is the equivalent of stating "2MASSW J0753321+291711 is very similar to Kelu-1 in the wavelength range 650 nm to 1000 nm." It is not the equivalent of stating "2MASSW J0753321+291711 has been measured to have $T_{\text{eff}} = 2000 \pm 50$ K and therefore we call it L2."[1] Once Kirkpatrick et al. (1999) defined nine types of L dwarfs from L0 to L8 to extend the previous M0–M9 system, the definition of those types in the optical has remained fixed, so that a circa 2000 classification should be the same as circa 2020 classification, but the interpretation of these spectra in terms of T_{eff} and any other physical parameters may well vary. It is generally accepted that intermediate half-spectral types such as L2.5—an appearance between L2 and L3 can be used—but more specific decimal numbers such as L2.1 are not a part of the system.

A spectral classification thus results by comparing an observed spectrum to the spectra of standard stars. Each paper defining a system also includes a text description of the most important features for each type, just as Cannon & Pickering (1901) did for the original OBAFGKM system. With our modern digital spectra, we can measure the strength of key atomic and molecular absorption features. Papers laying out the ultracool dwarf classification define specific numerical indices and typical ranges to guide the classifier, and in practice, an automated process like χ^2 minimization can also be very helpful to even a practiced human classifier. Classifying by direct comparison to a primary spectral standard may not be convenient or even necessary. There are many published, reliable classifications that can be used for "secondary" spectral standards. With modern digital data, observations can also be combined into a higher signal-to-noise template. The SDSS L2 template shown in Figure 3.1 was created by Schmidt et al. (2014) by combining seven different L2 dwarfs with the same spectroscopic setup: The SDSS templates are intended to be used both as a spectral type standard and a radial velocity standard for a survey that observed millions of stars and galaxies. The measured features could be used for automatic machine-learning algorithms.[2] We show the optical sequence for M6 to L2 dwarfs in Figure 3.2 and for L0 to L8 dwarfs in Figure 3.3 using the SDSS templates with the key atomic and molecular features marked.

Figure 3.2 shows the spectral sequence for M6-L1 dwarfs, illustrating the *M/L* transition. The strong TiO bands that characterize M stars begin to weaken in M7 dwarfs and the 740 nm band has nearly disappeared by L1. VO bands are strengthening in M7–M9 dwarfs and then begin to weaken. CrH and FeH strengthen throughout this range. These SDSS M dwarf templates have Hα (653 nm) in emission (designed by adding an "e" to the type, as in M9e) but many M dwarfs do not show Hα emission and it is not a classification criteria for

---

[1] Kelu-1 was later resolved into a binary system (Liu & Leggett 2005) consisting of an L2 and an L4, so that the spectrum shown in Figure 3.1 has a fainter L4 dwarf spectrum included in it, but as you can see from the figure, this makes little difference.

[2] We should note, however, a difference in philosophy here: The classifications are defined by a small number of primary standards, but machine-learning algorithms typically are designed to work with a very large training set.

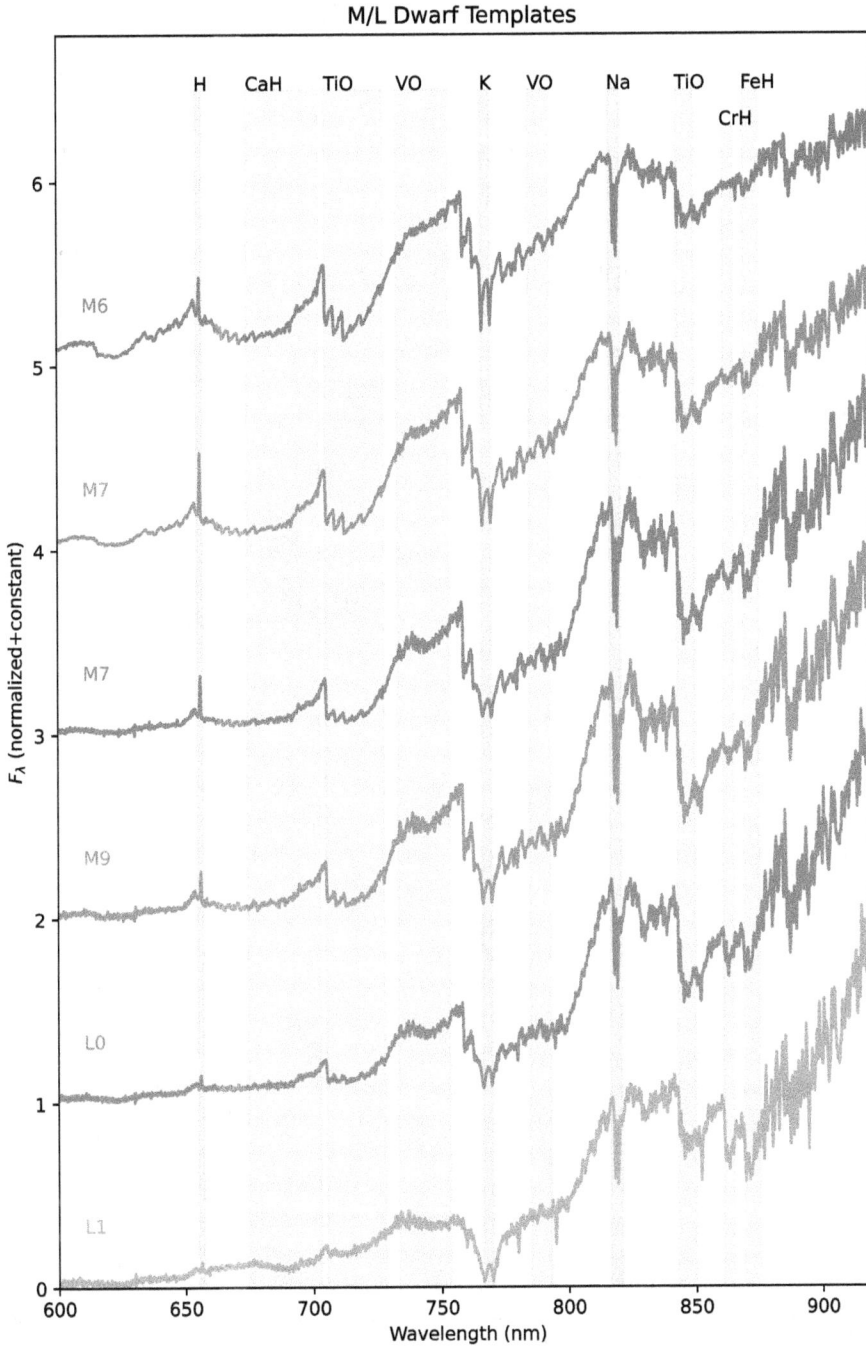

Figure 3.2. M and L dwarf templates formed by combining SDSS spectra of many individual objects can be used both for spectral type and for measuring radial velocities by cross-correlation. (Bochanski et al. 2007; Schmidt et al. 2014)

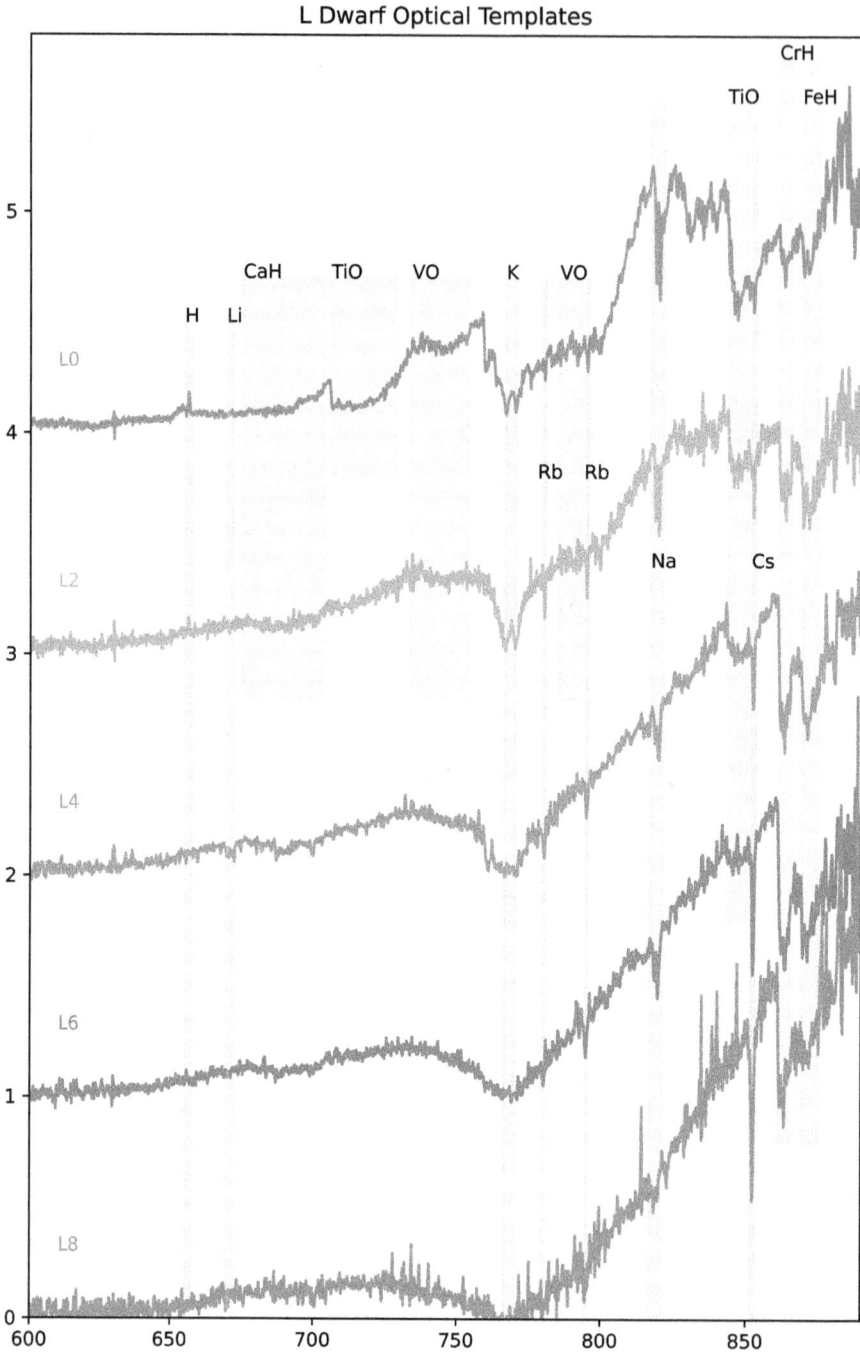

Figure 3.3. L dwarf templates created by averaging many previously classified L dwarfs together Schmidt et al. (2014). Note the trends in the molecular TiO, VO, VaH, CrH, and FeH. The alkali (Na, K, Cs, Rb) lines are The K doublet strengthens and widens until it becomes a broad feature at least ∼100 nm wide. (Noise is responsible for the apparent emission lines in the L8.)

M or L typing. The K doublet at 760 nm strengthens and broadens throughout this range. The L dwarf optical sequence is illustrated with the SDSS templates in Figure 3.3. Most of the molecular bands disappear. FeH strengthens then weakens. By L8, the overall red spectrum is dominated by the broadened K doublet feature, which becomes at least $\sim$100 nm wide, and bluewards of the region shown, the 589 nm Na doublet also becomes extremely broad. Throughout the sequence, the spectra becomes redder. Why are molecules like TiO and VO absent in the cooler L dwarfs? We may recall the AFGK stellar sequence, where the optical (Balmer) hydrogen absorption lines weaken and disappear. In that case, there is plenty of neutral hydrogen present in the cooler stars, but very little hydrogen is excited into the $n = 2$ energy level. This is a very different situation: If TiO was present at the temperatures and pressures of the L dwarf photosphere, the molecules would be in appropriate energy states to produce optical absorption. Instead, TiO is absent in cooler L dwarfs because the Ti becomes part of other molecules that condense into solid or liquid particles; the progressive disappearance of molecular features in the L dwarf optical sequence is due to a sequence of condensation of different materials. We will consider this process in our discussion of modeling brown dwarf atmospheres (Chapter 6).

Working in the near-infrared, near the peak of the spectral energy distribution, has great observational advantages, and researchers soon showed that even low resolution 0.9–2.4 μm spectra could be used to classify L dwarfs (Testi et al. 2001). Numerous research groups have investigated the near-infrared properties and classification schemes of L dwarfs. Geballe et al. (2002) divided the optical L8 type into an L8 and an L9 infrared type, but for the most part infrared systems have attempted to match the optical. One challenge, however, is illustrated in Figure 3.4,

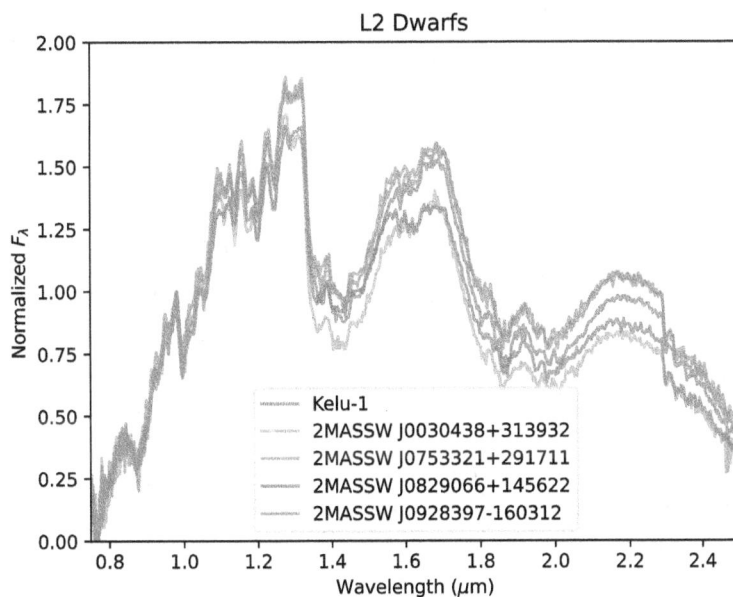

Figure 3.4. The same objects classified as L2 dwarfs in the optical as in Figure 3.1. All observations are from the SpeX Prism Library and analyzed with the SpeX Prism Library Analysis Toolkit (SPLAT) (Burgasser & Splat Development Team 2017).

which shows the same five L2 dwarfs as in Figure 3.1. Each spectrum is normalized to match in the "optical region." It is clear that the overall spectral slope from 1 to 2.4 μm, or $J - K$ color, varies among the objects even if most spectral features are similar. In other words, the one-parameter system that works so well in the optical is affected by (at least) one other parameter. Physically, this "second parameter" is mainly related to condensate cloud properties: thicker clouds make L dwarfs redder in the near-infrared. We've already noted that the optical spectra suggest that condensates are forming, though we will need a theory of cloud formation and structure to interpret the spectra. Kirkpatrick et al. (2010) defined new L dwarf standards in the infrared, chosen to have typical J–K colors, including an L9 standard. SpeX Prism spectra of these standards are shown in Figure 3.5. Figure 3.6 shows the same spectra overlaid on each other. On the other hand, Cruz et al. (2018) advocate for normalizing and comparing the J, H, and K windows separately in part to remove the effects of clouds. In any case, we can see the near-infrared spectra of L dwarfs show strong H_2O bands. The CO band is obvious at 2.3 μm. FeH at 0.99 μm becomes prominent but then disappears by L8. L9 shows very weak CH_4. Many more molecular and even atomic features can detected and analyzed in higher resolution near-infrared spectra: Compare the wiggle due to FeH at 1.22μm in these $R \approx 100$ spectra to the $R \approx 20\,000$ spectra shown in Figure 2.4.

The SpeX Prism Library Analysis Toolkit The (splat) Python package (Burgasser & Splat Development Team 2017) provides numerous tools to access and analyze over 2500 M, L, and T dwarf spectra collected in the SpeX Prism Library (Burgasser 2014). It is available at https://github.com/aburgasser/splat. The SpeX instrument is described by Rayner et al. (2003) and the extraction and calibration pipeline (spextool) is described by Cushing et al. (2004).

The STAR-LIKE paradigm predicts at least two types of ultracool dwarfs that will be spectroscopically peculiar: We expect a population of recently formed young (Population I, Galactic thin disk) objects with lower surface gravity (Figure 2.5) and a population of old, metal-poor Population II (thick disk and halo) objects (Figure 2.6). As increasing numbers of such objects have been identified, new classification systems to handle them have been proposed and widely adopted. Many low-gravity objects are members of moving groups or star clusters so that we can be confident in the physical interpretation.

For the low-gravity objects, Cruz et al. (2009) presented a system to classify intermediate gravity (β) and very low gravity (γ); the ordinary field L dwarf sequence is type α but this is not written. An example of an L3γ spectrum is shown in Figure 3.7. The atomic alkali lines (Na, K, Cs, Rb) are weaker than in the L3 standard, but many of the other features do not look dramatically different. (The system does not yet extend beyond L5, partly because sufficient examples are not yet observed, and partly because as we can see for the field L6 and L8 dwarfs, the alkali

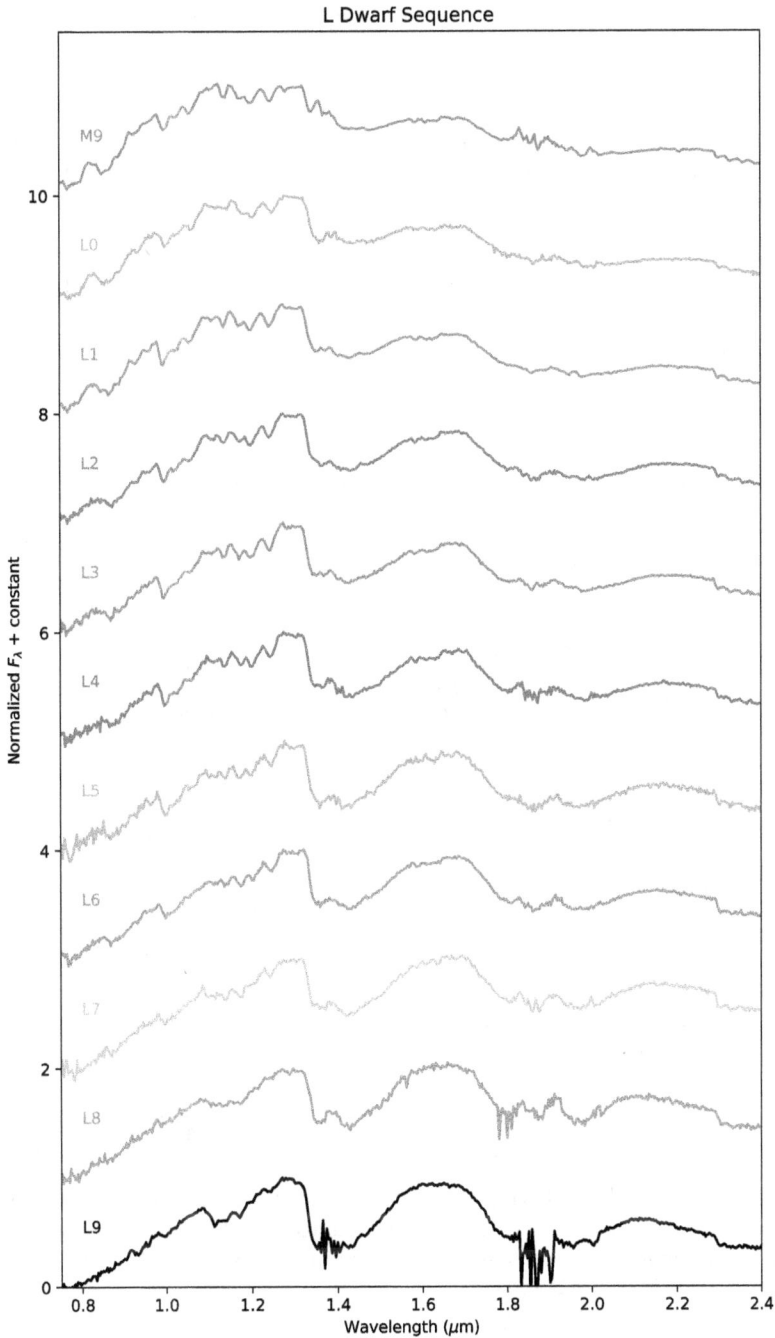

Figure 3.5. An infrared sequence of L dwarf standards as defined by Kirkpatrick et al. (2010). There is wide agreement among different groups that the broad features due to water and carbon monoxide (2.3μm) are correlated with optical spectra type.

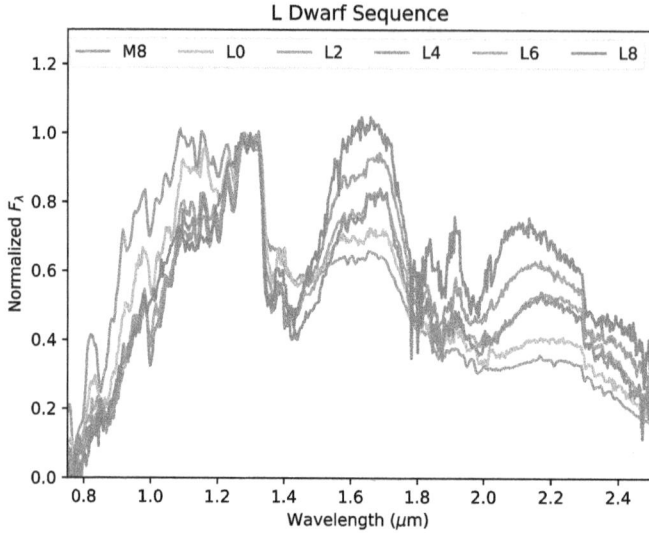

Figure 3.6. The L dwarf sequence overlaid. All objects are normalized to have the same flux density at 1.3 μm.

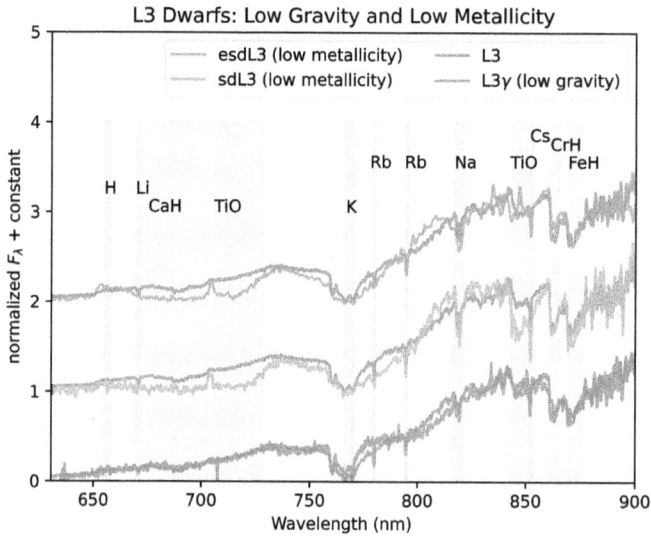

Figure 3.7. Examples of Peculiar L Dwarfs: 2MASS J22081363+2921215, a brown dwarf classified as a low-surface gravity L3γ (Cruz et al. 2009) and VVV J12564163-6202039, an sdL3 halo star (Zhang et al. 2019). The original sdL3 spectrum has been smoothed to lower resolution. Also shown is the L3 V standard 2MASS J11463449+2230527 (Kirkpatrick et al. 1999).

lines are the major features so there is little to compare to!) For the near-infrared, we can look at two different systems. Allers & Liu (2013) define spectral type standards for intermediate gravity types INT-G and very-low-gravity VL-G, shown in Figure 3.8. The Vl-G standards were all classified as γ type in the optical.

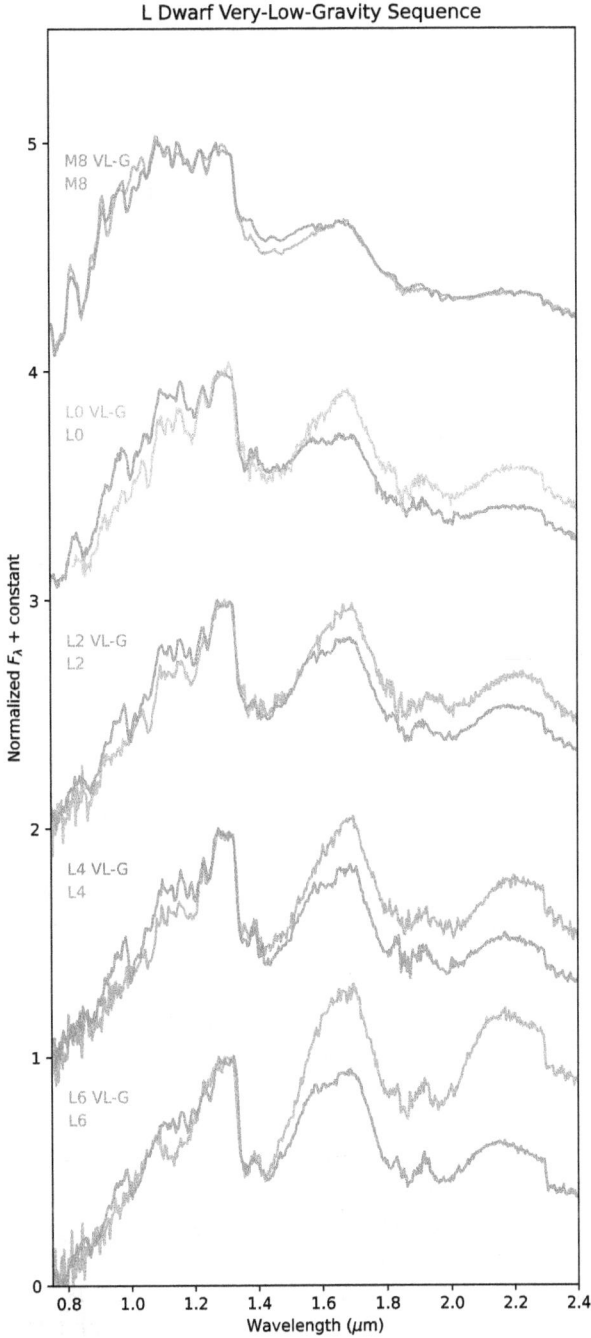

Figure 3.8. A very-low-gravity (VL-G) set of spectral standards (Allers & Liu 2013). All have also been classified as very-low-gravity γ in the optical system (Kirkpatrick et al. 2008). Note the peaked spectrum around 1.6 μm. These objects are also all redder than their field standard counterparts, and the spectral slope is used in the classification.

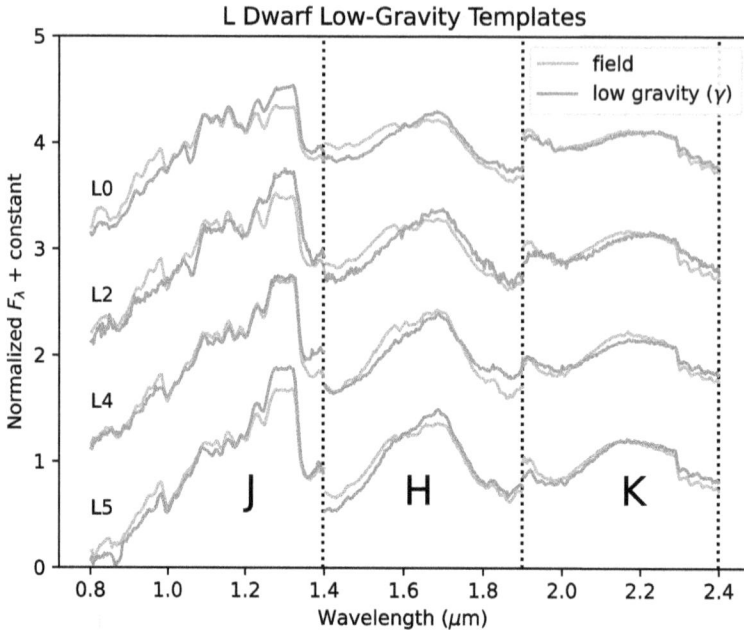

Figure 3.9. A set of template low-gravity dwarfs (Cruz et al. 2018). They recommend that each spectral region (J, H, K) be normalized separately to reduce the influence of color. Note again the peaked 1.6 μm region.

The low gravity objects are redder than the ordinary field L dwarfs and the H-band region looks sharply peaked compared to field L dwarfs. Cruz et al. (2018), as already noted, advocate for not using the red slope as a classification criteria and present templates for the field L dwarf and γ dwarf sequence. Roughly speaking, β or INT-G corresponds to ages of $\sim$100 Myr and γ or VL-G is more like $\sim$10 Myr (Figure 3.9).

The other expected class of peculiar ultracool dwarfs are the very-low-mass counterparts to metal-poor [Fe/H] $\lesssim$ 0.5, α element enhanced, stellar populations. These objects are called "subdwarfs."[3] Metal-poor M and L dwarfs can be identified by their spectral peculiarities, and a two-dimensional classification is needed because the range of metal abundances varies for orders of magnitudes. The M subdwarf system thus classifies normal (implicitly dM), subdwarf (sdM), extreme subdwarf (esdM), and ultra subdwarf (usdm) (Gizis 1997; Lépine et al. 2007), with the more metal-poor subdwarfs having weaker TiO absorption compared to CaH. This system has been extended to metal-poor L dwarfs: Zhang et al. (2017) identify sdL, esdL, and esdL. An example of an sdL3 and an esdL3 is shown in Figure 3.7. Like the M subdwarfs, L subdwarfs have strong atomic alkali lines (Na, K, Cs, Rb)

---

[3] A warning: This term could be considered misleading. Although subgiants are much smaller than giants, subdwarfs have peculiar spectra and positions in the HR diagram due to their low metallicity, not a small radius.

and CaH. However, in contrast, the sdL3 and esdL3 subdwarfs show TiO where it is absent in the solar metallicity L3 standard, presumably a sign of different condensation efficiency in metal-poor atmospheres. Zhang et al. (2017) estimate that the M subdwarf and L subdwarf classifications are roughly consistent in metallicity, with sdL having $-0.3 \gtrsim$ [Fe/H] $\gtrsim -1.0$, esdL $-1.0 \gtrsim$ [Fe/H] $\gtrsim -1.7$, and usdL [Fe/H] $\lesssim -1.7$. The L subdwarfs are bluer in the near-infrared. A third cause of peculiar classifications is the presence of an unresolved binary. An unresolved T dwarf companion, for example, may add a weak methane feature to a late-M or L dwarf primary. Bardalez Gagliuffi et al. (2014) describe a system for identifying such systems from low-resolution near-infrared spectra.

Spectroscopy in the mid-infrared can complement the optical and near-infrared. Here, the emphasis is more on physical diagnosis than creating a new shared language of spectral types. In Figure 3.10, we show some examples of 5–14 μm $R \approx 90$ spectra observed with the Spitzer Space Telescope. The presence of a broad silicate feature, which is correlated with $J - K$ color in mid-L dwarfs, supports the cloud model for understanding L dwarfs. Furthermore, both $J - K$ color and the silicate feature are correlated with the viewing angle ($\sin i$) inferred from variability studies, suggesting that the equatorial regions have thicker clouds than the polar regions (Vos et al. 2020; Suárez et al. 2023). If so, the simple approximation that T_{eff}, $\log g$, and the elemental abundances determine the low-resolution appearance has notable limitations.

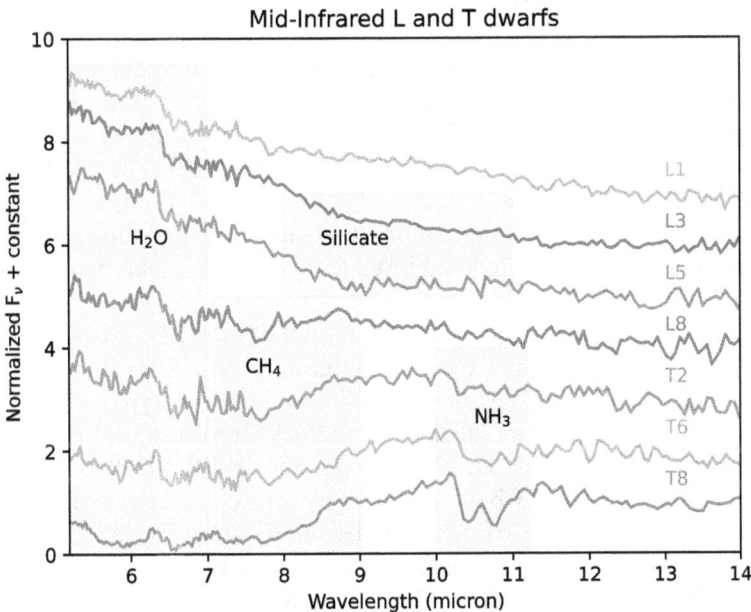

Figure 3.10. Selected L and T dwarfs from Suárez & Metchev (2022)'s uniform analysis of all ultracool dwarfs observed with the Spitzer Space Telescope IRS spectrograph. Note the broad silicate feature in mid-L dwarfs.

3.3 T and Y Dwarfs

Following the initial spectroscopic exploration of Gliese 299B and other dwarfs with methane absorption, T dwarf spectral standards for the near-infrared were defined by Burgasser et al. (2006) and are plotted in Figure 3.11. Deep H_2O absorption bands are present throughout the sequence, and methane appears in the H and K regions. The position of the molecular band for water and methane used by Burgasser et al. (2006) is shown with the standards overlaid and (arbitrarily) normalized at 1.3 μm in Figure 3.12. Overall, the spectra become bluer, that is, the H and K regions are fainter than the J region. It should be clear that each type is distinct even at these low resolutions. Again, we must not forget that the low resolution blurs together what would be individual features at high resolution. Examine Figure 3.13, which shows the highest resolution H band spectrum ever published for a T6 dwarf (Tannock et al. 2022): The smooth regions in our low-resolution figure are actually a series of sharp features, and in the detailed plot for 1.5800μ m $< \lambda < 1.6000 \mu$ m, the peak of the spectral energy distribution in the H band, numerous sharp H_2O lines can be seen along with a line that Tannock et al. (2022) show is the first detection of H_2 in a brown dwarf. There is no line-free continuum in ultracool dwarfs. In principle, spectral typing for T dwarfs can also be done in the optical. Burgasser et al. (2003) describe the appearance of red optical spectra for T dwarfs. Atomic Na, Ca, K, Rb and molecular CH_4, CrH, FeH, and H_2O are all noted. In practice, however, optical spectral types for T dwarfs require long exposure times on large telescopes and become increasingly impractical as T dwarfs get intrinsically fainter.

Just as with the M and L sequence, we expect both young, low gravity and old, metal-poor populations for T dwarfs. The intrinsic faintness of these objects, however, means that relatively few are known and so there are not yet well-defined classification systems. 2MASS J13243553+6358281, for example, is described as a red T2 dwarf that is likely ~150 Myr old (Gagné et al. 2018). The red colors may be explained by weaker collision-induced H_2 opacity and thicker clouds. Similar spectra are seen in T dwarf companions to young stars. Similarly, the sdT and esdT regime is just being explored and just a few examples are known (Meisner et al. 2023).

Finally, we turn to the objects beyond T8. UGPS J072227.51-054031.2 (Lucas et al. 2010), clearly cooler than the T8 standard, is defined as T9. Cushing et al. (2011) described WISEP J182831.08+265037.8 as the "archetypal member of the Y class" and defined WISEP J173835.52+273258.9 as Y0, and identified several similar objects. Kirkpatrick et al. (2012) suggested WISE J035000.32-565830.2 as a tentative Y1 standard. The J-band flux region narrows, and the relative strength of H increases—that is, Y dwarfs are redder in $J - H$. For WISEP J182831.08 +265037.8, the height of the F_λ peaks is similar for the J and H bands. On the other hand, WISE J085510.83-071442.5 (Luhman 2014) is clearly cooler than the other early Y dwarf discoveries. We put it on the plots later in this chapter as Y4 simply for the purposes of visualization. Observations of these cool sources are intrinsically difficult from the ground. Many Y dwarfs were observed in the first year

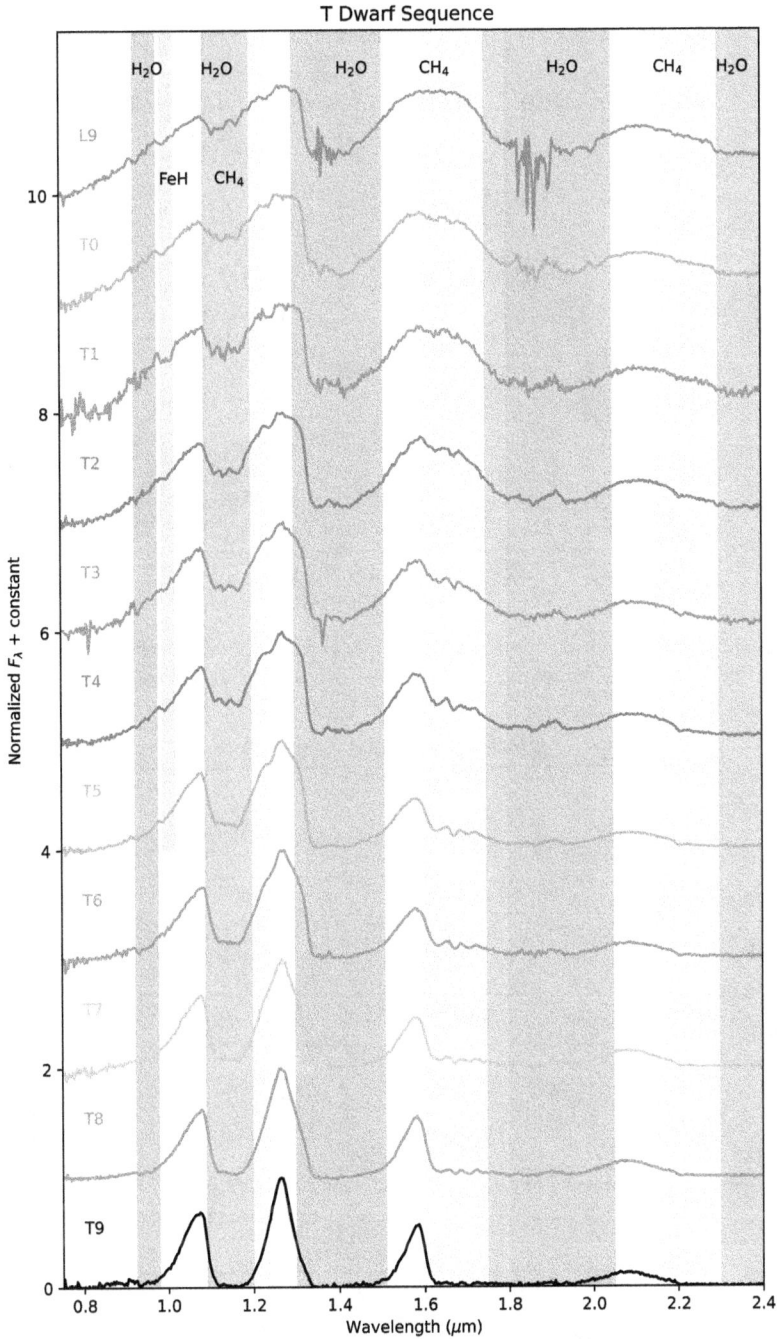

Figure 3.11. T dwarfs spectral standards as observed with the IRTF Spex Prism mode. The Y0 spectrum is from JWST (Beiler et al. 2023). All objects are normalized to have the same flux density at 1.3 μm. Some prominent absorption features are identified by the shaded regions.

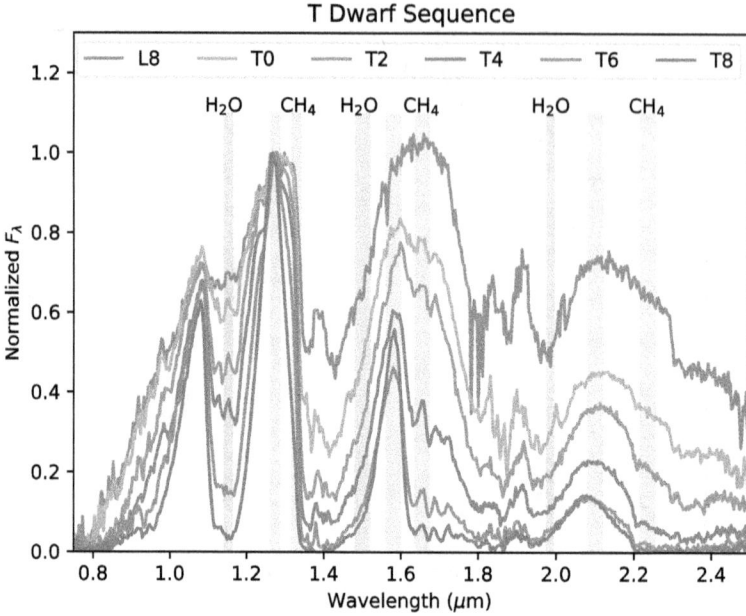

Figure 3.12. The T dwarf standards overplotted and normalized to have the same flux density at 1.3 μm. The wavelength regions of spectral indices used in the classification scheme are shaded.

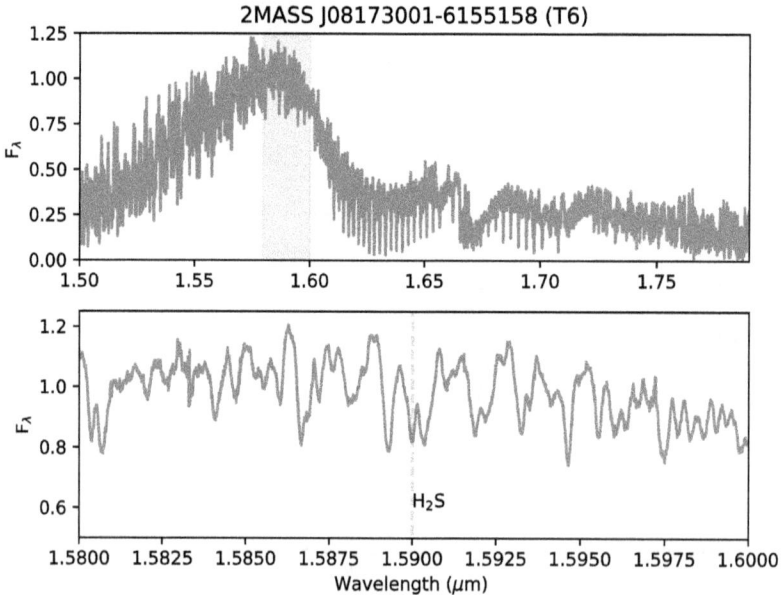

Figure 3.13. An $R \approx 45\,000$ H-band spectrum of a T6 dwarf (Tannock et al. 2022) with $V \sin i = 22.5 \pm 0.9$ km s^{-1} and $v_{\text{rad}} = 6.1 \pm 0.5$ km s^{-1}. Note the many sharp features compared to the low-resolution classification spectra: The signal-to-noise ratio is >200 and all features are real. The lower plot shows the region 1.58 μm to 1.60 μm. Nearly every absorption line in this region is due to H$_2$O but (Tannock et al. 2022) demonstrates that the 1.5900 μm line is due to H$_2$S.

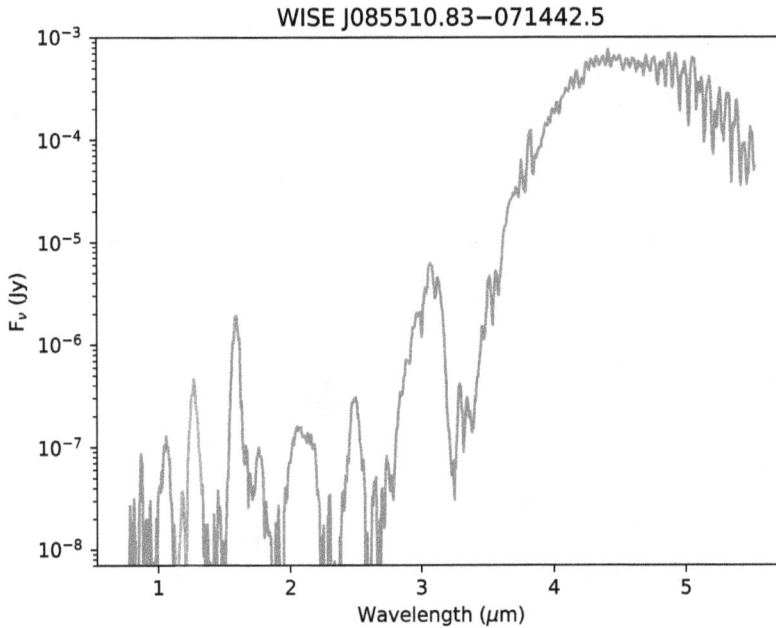

Figure 3.14. JWST spectrum of the coolest known brown dwarf, WISE J085510.83-071442.5 (Luhman et al. 2024). Note how little flux is observed in the near-infrared wavelength region most accessible from the ground (<2.6 μ m).

of JWST operations and the spectroscopic results are just beginning to be published at this writing in July 2024. It is plain that the 2011–2023 work on Y dwarfs will be quickly superseded by JWST high signal-to-noise spectra from 1 to 12 μm. The first such Y0 spectrum is shown in Figure 3.7. The JWST1–5 μm spectrum (Figure 3.14) of W0855-0714 was published by Luhman et al. (2024) but the longer wavelength data are not yet available. Beiler et al. (2024) observed 23 T and Y dwarfs with JWST from 1 to 12 μm. They conclude that "the near-infrared spectral type ordering does not map directly to effective temperature." In the region of peak flux, "the 5 μm region does not follow a temperature sequence, nor could we identify one by eye. This lack of a sequence is especially true among the Y dwarfs." For the 8–12 μm region, "the spectral morphology of this wavelength range appears to be the most strongly correlated with effective temperature." As they note, all this suggests that the $\log g$ and elemental abundances likely play a major (rather than secondary) role in the appearance of Y dwarf spectra, breaking the one-parameter assumption that worked so well for L and T dwarfs. We might even speculate that SUPER-JUPITER scenarios could matter in some of these planetary mass objects. Nevertheless, this only reinforces the need for an empirical system to describe Y dwarf spectra as theoretical models are developed. Many more Y dwarfs are being observed with JWST. In Chapter 6, we discuss more of the atmospheric modeling that is needed to understand Y dwarfs, and we must view this as field in the midst of a revolution.

3.4 Luminosities and Effective Temperatures

Within the many limits already noted, we expect spectral types to be primarily determined by effective temperature even if that assumption is now called into question for Y dwarfs. We also expect on theoretical grounds (review Figure 1.2) that the brown dwarf cooling sequence is rather narrow in the theoretical H–R diagram (effective temperature and luminosity.) It is therefore possible to follow the approach outlined in Chapter 2 and calculate an ultracool dwarf's luminosity with Equation (2.2) for hundreds of nearby ultracool dwarfs: The distance is calculated from the observed parallax and the flux is calculated by integrating the observed spectral energy distribution. One can then in turn estimate the effective temperature with Equation (2.8) by adopting an expected radius from models. In Figures 3.15 and 3.16, we show the fitted polynomial relationship for the regular ultracool M, L, and T dwarf sequence determined by Sanghi et al. (2023), which are similar to those determined earlier by Filippazzo et al. (2015). Always remember, however, that any spectral type relationship for the normal field dwarfs by definition does not apply to the various types of peculiar, low-gravity, or low-metallicity spectral types. The reason the L3, L3γ, sdL3, and esdL3 spectra shown in Figure 3.7 all have "L3" in their classification is simple; they have some resemblances and it was convenient to use the same number, not that there is any expectation that they have same effective temperature or luminosity. The peculiar types must be treated separately.

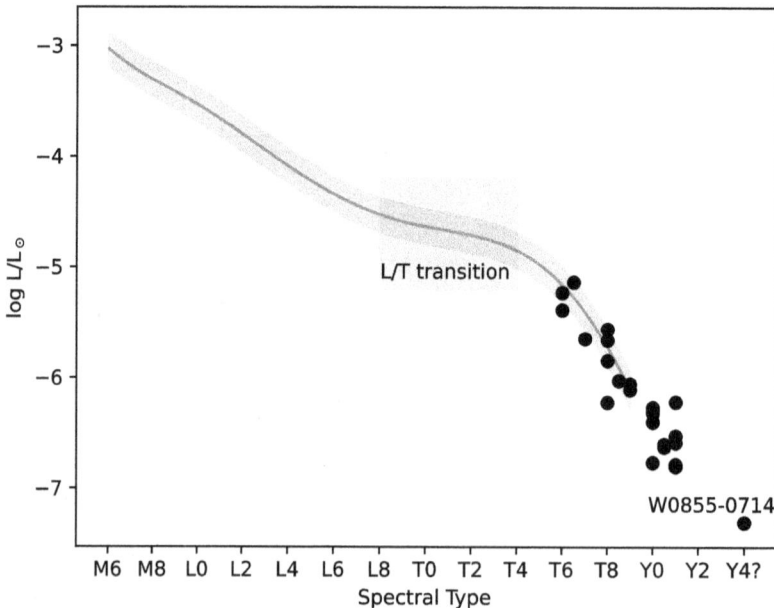

Figure 3.15. The fitted relationship between spectral type and luminosity as measured for a large sample by Sanghi et al. (2023). The observed scatter in log $\mathcal{L}$ is ±0.157. The black data points show individual cool objects measured with JWST (Beiler et al. 2024). W0855-714 (Luhman et al. 2024) is plotted as Y4, but the system beyond Y1 is not yet defined.

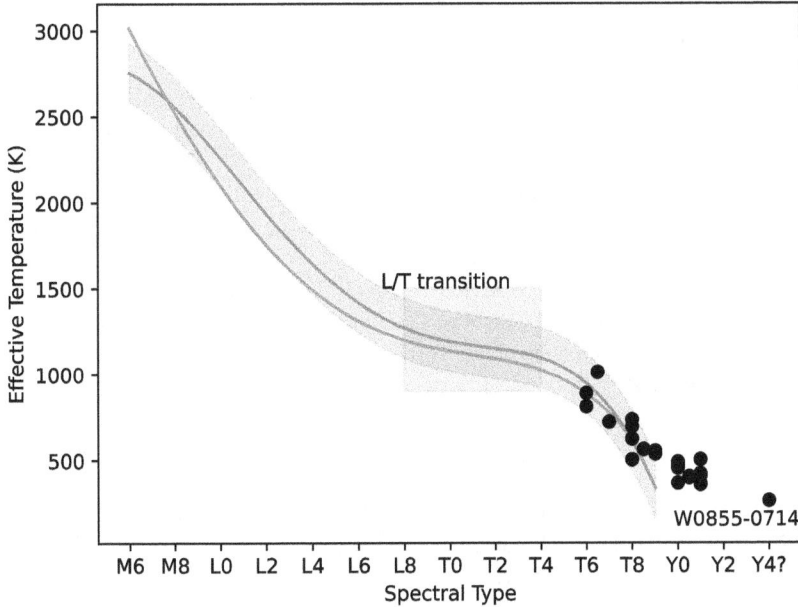

Figure 3.16. The fitted relationship between spectral type and effective temperature as estimated from measured luminosities and model radius predictions for a large sample by Sanghi et al. (2023). The observed scatter in $\log \mathcal{L}$ is ±175K. The black data points show individual cool objects measured with JWST (Beiler et al. 2024). W0855-714 (Luhman et al. 2024) is plotted as Y4, but the system beyond Y1 is not yet defined. The fitted relationship for a sample of low-gravity dwarfs is also shown to illustrate the fact that the physical properties for peculiar spectral types will be different.

Figure 3.16 also shows the fitted polynomial relationships for objects classified as low gravity: Note that low-gravity M dwarfs are hotter than their field counterparts, but low-gravity L dwarfs are cooler than the field sequence.

These empirical luminosities and model-dependent effective temperatures confirm the success of the spectral type approach. The spectral type sequence is primarily an effective temperature approach and the observational HR diagram (luminosity versus spectral type) is consistent with the theoretical prediction of a brown dwarf cooling sequence. However, there is a surprising result: Although it is quite obvious from the spectra shown in Figures 3.11 and 3.12 that the spectra look distinct and form a clear sequence as we proceed from L8 to T5, this large "spectral type" range corresponds to a narrow range in luminosity and effective temperature. Understanding this "L/T transition" has been a key challenge for brown dwarf research. Also shown in Figures 3.15 and 3.16 are the individual luminosities and effective temperatures (360–500 K) determined with this technique from the JWST late-T and Y dwarf observations by Beiler et al. (2024) plus the cooler W0855-714 (Luhman et al. 2024). The result confirms that these objects are, on average, less luminous and cooler than T dwarfs, even if the scatter confirms the need for future observations and theoretical work.

References

Allers, K. N., & Liu, M. C. 2013, ApJ, 772, 79

Bardalez Gagliuffi, D. C., Burgasser, A. J., Gelino, C. R., et al. 2014, ApJ, 794, 143

Beiler, S. A., Cushing, M. C., Kirkpatrick, J. D., et al. 2023, ApJL, 951, L48

Beiler, S. A., Cushing, M. C., Kirkpatrick, J. D., et al. 2024, ApJ, 973, 107

Bochanski, J. J., West, A. A., Hawley, S. L., & Covey, K. R. 2007, AJ, 133, 531

Burgasser, A. J. 2014, Astronomical Society of India Conf. Series, Vol. 11, (Banglalure: Astronomical Society of India) Conf. Series, 7

Burgasser, A. J., Geballe, T. R., Leggett, S. K., Kirkpatrick, J. D., & Golimowski, D. A. 2006, ApJ, 637, 1067

Burgasser, A. J., Kirkpatrick, J. D., Liebert, J., & Burrows, A. 2003, ApJ, 594, 510

Burgasser, A. J.Splat Development Team 2017, Astronomical Society of India Conf. Series, Vol. 14, (Banglalure: Astronomical Society of India) Conf. Series, 7

Cannon, A. J., & Pickering, E. C. 1901, AnHar, 28, 129

Cruz, K. L., Kirkpatrick, J. D., & Burgasser, A. J. 2009, AJ, 137, arxiv:0812.0364

Cruz, K. L., Núñez, A., Burgasser, A. J., et al. 2018, AJ, 155, 34

Cushing, M. C., Vacca, W. D., & Rayner, J. T. 2004, PASP, 116, 362

Cushing, M. C., Kirkpatrick, J. D., Gelino, C. R., et al. 2011, ApJ, 743, 50

Filippazzo, J. C., Rice, E. L., Faherty, J., et al. 2015, ApJ, 810, 158

Gagné, J., Allers, K. N., Theissen, C. A., et al. 2018, ApJL, 854, L27

Geballe, T. R., Knapp, G. R., Leggett, S. K., et al. 2002, ApJ, 564, 466

Gizis, J. E. 1997, AJ, 113, 806

Kirkpatrick, J. D., Henry, T. J., & McCarthy, D. W. Jr+ 1991, ApJS, 77, 417

Kirkpatrick, J. D., Reid, I. N., Liebert, J., et al. 1999, ApJ, 519, 802

Kirkpatrick, J. D., Reid, I. N., Liebert, J., et al. 2000, AJ, 120, 447

Kirkpatrick, J. D., Cruz, K. L., Barman, T. S., et al. 2008, ApJ, 689, 1295 arxiv:0808.3153

Kirkpatrick, J. D., Looper, D. L., Burgasser, A. J., et al. 2010, ApJS, 190, 100

Kirkpatrick, J. D., Gelino, C. R., Cushing, M. C., et al. 2012, ApJ, 753, 156

Lépine, S., Rich, R. M., & Shara, M. M. 2007, ApJ, 669, 1235

Liu, M. C., & Leggett, S. K. 2005, ApJ, 634, 616

Lucas, P. W., Tinney, C. G., Burningham, B., et al. 2010, MNRAS, 408, L56

Luhman, K. L. 2014, ApJL, 786, L18

Luhman, K. L., Tremblin, P., Alves de Oliveira, C., et al. 2024, AJ, 167, 5

Martín, E. L., Delfosse, X., Basri, G., et al. 1999, AJ, 118, 2466

Meisner, A. M., Leggett, S. K., Logsdon, S. E., et al. 2023, AJ, 166, 57

Morgan, W. W., & Keenan, P. C. 1973, ARA&A, 11, 29

Rayner, J. T., Toomey, D. W., Onaka, P. M., et al. 2003, PASP, 115, 362

Sanghi, A., Liu, M. C., Best, W. M. J., et al. 2023, ApJ, 959, 63

Schmidt, S. J., West, A. A., Bochanski, J. J., Hawley, S. L., & Kielty, C. 2014, PASP, 126, 642

Suárez, G., & Metchev, S. 2022, MNRAS, 513, 5701

Suárez, G., Vos, J. M., Metchev, S., Faherty, J. K., & Cruz, K. 2023, ApJL, 954, L6

Tannock, M. E., Metchev, S., Hood, C. E., et al. 2022, MNRAS, 514, 3160

Testi, L., D'Antona, F., Ghinassi, F., et al. 2001, ApJL, 552, L147

Vos, J. M., Biller, B. A., Allers, K. N., et al. 2020, AJ, 160, 38

Zhang, Z. H., Burgasser, A. J., & Smith, L. C. 2019, MNRAS, 486, 1840

Zhang, Z. H., Pinfield, D. J., Gálvez-Ortiz, M. C., et al. 2017, MNRAS, 464, 3040

An Introduction to Brown Dwarfs
From very-low-mass stars to super-Jupiters
John Gizis

Chapter 4

Photometry and Astrometry

4.1 Photometry

For our purposes, photometry is defined as measuring the brightness of stars using digital images taken through well-defined filters. "Optical" measurements for $300\text{nm} \lesssim \lambda \lesssim 1000$ nm use silicon-based charge-coupled devices and "infrared" use other detector technologies, but both have high quantum efficiency to convert incident photons into measurable counts. We have already seen the magnitude system and now will explore magnitudes in more detail. Figure 4.1 shows the red optical (*grizy*) Pan-STARRS and Gaia (R_P) transmission response as a function of wavelength. Figure 4.2 shows the 2MASS and Mauna Kea Observatory (MKO) ground-based near-infrared (*JHK*) filters[1] and the 3–5 μm WISE and Spitzer Space Telescope filters. Not shown are the large number of HST and JWST filters that together cover the range from the ultraviolet through the mid-infrared. Each measurement reported for a star is a magnitude in that system. For example, the L1 dwarf WISEP J190648.47+401106.8 (W1906+4011) has *Gaia* $G = 17.839 \pm 0.003$, Gaia $R_P = 16.245 \pm 0.005$, Pan-STARRS (PS1) $i = 17.420 \pm 0.005$, 2MASS $J = 13.078 \pm 0.024$, and 2MASS $K = 11.771 \pm 0.005$.

What do these magnitudes mean? From the purely observational point of view, apparent magnitudes are measurements of the relative count rate of different stars. Let us use the 2MASS J magnitude for WISEP J190648.47+401106.8 (2MASS J19064801+4011089) as a case study. 2MASS is on a Vega system where the zero-point of the magnitude scale in all filters is defined by Vega, so that $J = 13.078$ means W1906+40 is nominally $10^{-13.078/2.5} = 5.782 \times 10^{-6}$ times fainter than Vega in the J filter. The purpose of 2MASS was to measure this photometry for stars over the entire celestial sky.

---

[1] The 2MASS K filter is officially called a *Ks* (K-short) filter.

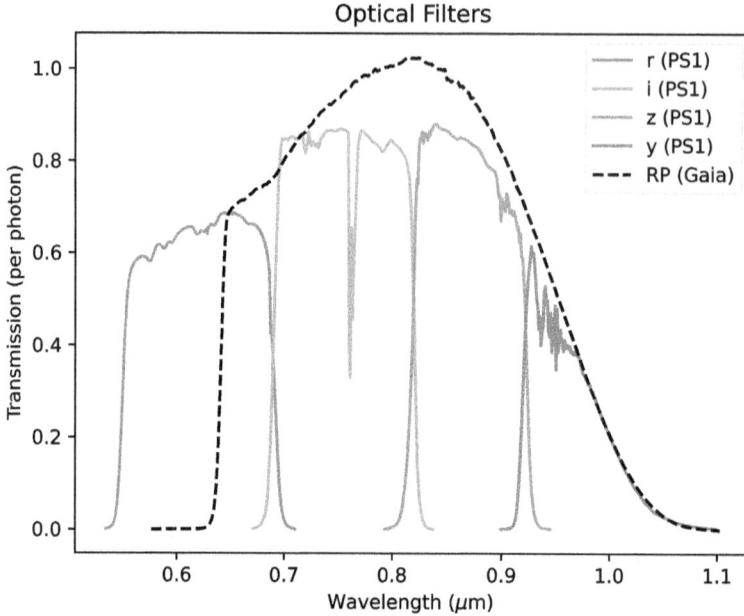

Figure 4.1. Transmission curves for selected optical filters: Pan-STARRS (PS1) and the red Gaia filter. Note the telluric absorption in PS1 filters. These are photon counting responses, so that the λ/hc term in Equation (4.3) is needed if using a F_λ spectrum.

Filter Profiles Each observatory documents their own filter response profiles, but many have been collected by the Spanish Virtual Observatory (SVO) at https://svo.cab.inta-csic.es along with the corresponding absolute calibration zero-points. The SVO also has an extensive collection of theoretical spectra and evolutionary models. The filter profiles may be used with Gaia spectra to synthesize optical photometry in any desired filter Gaia Collaboration et al. (2023)

The 2MASS telescopes scanned each field so that each photometric measurement reported for a star is a result of of six separate images taken consecutively with a total exposure time of 7.8 s. The count rate (counts per second) for each star in a field was transformed into an instrumental magnitude by the 2MASS pipeline. This count rate depends on the intrinsic brightness (F_λ) of the star, but also the transmission of the filter, atmosphere, telescope and camera optics, and detector response. These are wavelength dependent, but can also slowly depend on time, and of course each observation had its own path length through the atmosphere. These effects need to be removed. If we write the airmass (x) as the secant of the zenith angle, the instrumental magnitude was calibrated with a nightly zero-point and an extinction correction (k):

$$J = J_{\text{inst}} - k_J(x - 1) + \text{ZP}_J \tag{4.1}$$

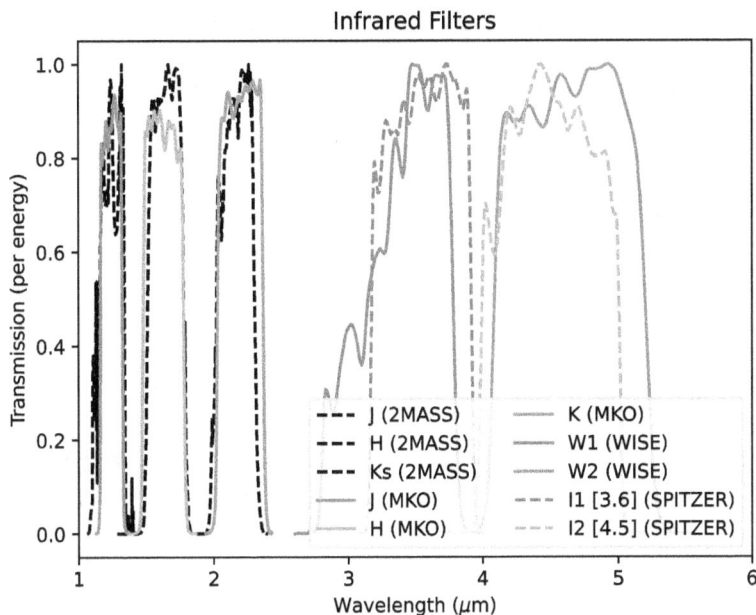

Figure 4.2. Transmission curves for selected infrared filters: Note the differences between the 2MASS and MKO responses, such as the red cutoff in the 2MASS K "short" filter and the difference between WISE and Spitzer. These are energy counting responses, that is, the λ/hc term in Equation (4.3) has been included in the transmission f_λ.

For 2MASS, the nightly zero-points (ZP_J) and monthly airmass corrections (k_J) were determined by observations of special calibrator fields every 2 h. The 2MASS system is uniform around the sky at the $\sim$1% level and is not transformed in any way to match other near-infrared filter measurements other than selecting the overall calibration star value (Nikolaev et al. 2000). Vega itself is too bright to serve as a calibrator star.[2] Rather, the 2MASS system was chosen to agree on average with a set of previously published $J \approx 11$ (Persson et al. 1998) stars, which themselves were previously calibrated relative to a group of $J \approx 7$ standards (Elias et al. 1982), themselves calibrated relative to $J \approx 3 - 4$ standards, finally tied to Vega itself. Each of these steps was thought to be accurate at the 1% level, but were taken with different filters and detector technologies, so we do not know that Vega would literally be $J = 0.000$, $H = 0.000$, $K = 0.000$ if 2MASS could have observed it reliably. The 2MASS magnitude system is thus very well defined as its own system, but its absolute calibration and relationship to other near-infrared systems needs further discussion.

---

[2] Besides its extreme brightness, Vega has at least two other problems in practice. First, it is viewed pole-on which makes its spectrum anomalous compared to most other A stars and correspondingly difficult to model. Second, Vega has a debris disk which makes it appear brighter than its photosphere alone at wavelengths greater than 10 μm. There have been great efforts in recent years to improve the optical and infrared flux calibration system.

This approach of applying a zero-point and airmass correction is standard for Earth-based observatories. If the nightly measurements of calibrators give inconsistent zero-points, then the cause is likely clouds and the night is considered non-photometric. Here, we can see the advantage of space-based observatories: Without the atmosphere, the airmass correction is zero, we never have cloudy nights, and the zero-points are very stable over the long term. However, when making your own targeted observations, it may not longer be necessary to observe enough calibrators to solve for k and ZP: The 2MASS stars in the field of view can be used to set the zero-point. This approach was used in later, deeper near-infrared sky surveys such as the UKIRT Deep Sky Survey (Hodgkin et al. 2009), although the calibration procedure had to account for the differences between the 2MASS J, H, K filters and the UKIRT MKO ones. The situation in the optical is parallel to the near-infrared one. The optical PS1 survey was able to determine nightly zero-points and extinction corrections for each of its five filters (g, r, i, z, y) so that the entire system is consistent (Schlafly et al. 2012), and the overall zero-points are based on a separate analysis to place on an AB system (Tonry et al. 2012), which we explain later. Now, however, space-based Gaia photometry and spectroscopy is consistent over the entire sky and may in the future be able to calibrate all optical photometric systems (Gaia Collaboration et al. 2023). The Vera C. Rubin Observatory LSST will benefit from Gaia calibrator stars in the field of view and simultaneous spectroscopic monitoring of the telluric absorption. In any case, we are currently in an excellent position to make empirical statements like VB10 (2MASS $J = 9.908 \pm 0.025$, Gaia $G = 14.303 \pm 0.003$) is 18.5 times brighter than W1906+4011 in the 2MASS J filter system but 26.0 times brighter than W1906+40 in the Gaia G filter. This is useful, but we would also like to be able to physically interpret these measurements.

Photometry Software The `photutils` Python package (Bradley et al. 2024) includes tools to measure the count rate of stars from astronomical images. Both aperture and point-spread function (PSF) photometry can be used. https://photutils.readthedocs.io/

The observed count rate depends upon the star's spectral energy distribution as a function of wavelength: $F_\lambda(\lambda)$ with the units of energy per unit area per second per unit wavelength (such as ergs cm^{-2} s^{-1} Å^{-1} or W m^{-2} s^{-1} nm^{-1}). We observe the star through a filter and telescope optics—and the Earth's atmosphere if ground-based—with a transmission rate P_λ. The elements contributed to P_λ—the detector quantum efficiency, the transmission of the filter, etc.—can be measured in the laboratory beforehand or modeled in the case of the atmosphere. To get counts, we must divide F_λ by the energy per photon (hc/λ). For a Vega system, the corresponding magnitude is

$$m = -2.5 \log_{10}(C/C_{\text{Vega}}). \tag{4.2}$$

$$m = -2.5 \log_{10} \left(\frac{\int P_\lambda F_\lambda \lambda \, d\lambda}{\int P_\lambda F_{\lambda,\text{Vega}} \lambda \, d\lambda} \right) \qquad (4.3)$$

What is the physical meaning of saying $J = 13.078$ in terms of photons or watts? Cohen et al. (2003) model 2MASS and determine that Vega with $J = 0$ has a total in-band irradiance of 5.082×10^{-14} W cm^{-2} ($\pm 1.608\%$) with a bandwidth of 0.162 μm. This can also be expressed at an "isophotal" wavelength of $\lambda = 1.235$ μm and $F_\lambda = (3.129 \pm 0.055) \times 10^{-13}$ W cm^{-2} μm^{-1} or an isophotal frequency of $\nu = 2.428 \times 10^{14}$ Hz and $F_\nu = 1594 \pm 28$ Jy. Furthermore, even though this near-infrared calibration system uses a different set of secondary standards as the path from Vega to 2MASS, Cohen et al. (2003) report that Vega's estimated J-band agrees to 0.001 magnitudes. These zero-points can be used to convert photometric zero-points to F_λ or F_ν on spectral energy distribution plots. Similar zero-points are provided with the filters at the observatory sites or the Spanish Virtual Observatory service.

Using Vega for the zero-points means that the flux density zero-points are a strong function of wavelength and much lower in the near-infrared than the optical. The SDSS, PS1, and LSST *ugrizy* systems instead use AB zero-points. The monochromatic AB magnitude system (Oke 1974; Tonry et al. 2012) defines the zero-point to be $m_{\text{AB}} = -2.5 \log_{10} f_\nu - 48.60$ (f_ν in erg s^{-1} cm^{-2} Hz^{-1}), so that for broad filters:

$$m = -2.5 \log_{10} \left(\frac{\int P_\nu F_\nu (h\nu)^{-1} d\nu}{\int P_\nu (3631 \text{Jy})(h\nu)^{-1} d\nu} \right) \qquad (4.4)$$

HST uses the STMAG system which is instead defined for constant F_λ, so that $m_{\text{ST}} = -2.5 \log_{10} f_\lambda - 21.10$ (F_λ in erg s^{-1} cm^{-2} Å^{-1}.) The AB and STMAG zero-points were chosen to agree with Vega at 548 nm (V band).

In any case, ultracool dwarfs have a very different spectral energy distribution than Vega, other calibrator stars, or the hypothetical constant F_ν or F_λ sources. Figure 4.3 shows the count rate for Vega and W1906+4011, both normalized to 2MASS $J = 13.078$. Clearly, the W1906+4011 J-band detected photons are on average redder and therefore less energetic. It is therefore recommended, if possible, to use theoretical or observed spectra to synthesize magnitudes to compare to observations.

Synthetic Photometry synphot is a Python package that simulates HST photometry and spectroscopy but can also be used for other telescopes and filters (STScI Development Team 2018). It is described at https://synphot.readthedocs.io/.

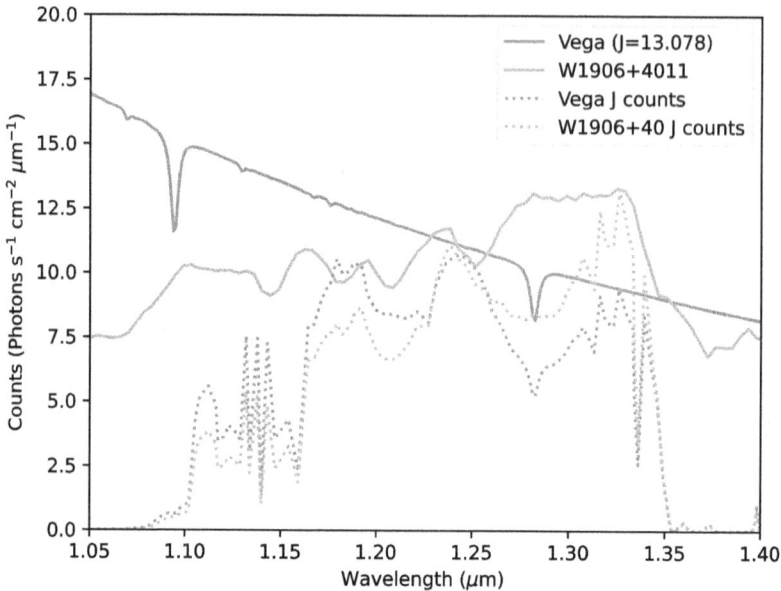

Figure 4.3. The photon spectrum for W1906+4011 and Vega, both normalized to have the same 2MASS magnitude $J = 13.078$ using synphot. The L dwarf is much different from the A0 dwarf, complicating interpretation in terms of the energy. Similar effects cause color terms between different filters, such as MKO and 2MASS J band.

The differences between ultracool dwarf spectra and typical calibration stars is also important when transforming between similar filters, such as different versions of the J filter. The MKO system is now the standard set of JHK filters at most ground-based observatories (Tokunaga et al. 2002; Simons & Tokunaga 2002). As seen in Figure 4.2, the MKO filter system is different from 2MASS. Accordingly, J, H, K magnitudes in MKO will be slightly different from the 2MASS ones, and these "color terms" will be a function of the star's color or spectral type. Stephens & Leggett (2004) discusses transformations between MKO and other JHK filter systems, including 2MASS, and the dependence on atmospheric conditions. Many J filters, including 2MASS, are affected by telluric water absorption which affects the calibration stars differently than the T dwarfs with intrinsic deep water absorption bands. The UltracoolSheet (Best et al. 2020) includes both 2MASS and MKO photometry when available.

---

[3] An extraordinary example of how over longer timescales we must also account for the radial velocity is the ultracool dwarf binary WISE J072003.20-084651.2 (Scholz's Star), which currently has a proper motion of $\sim$0.1 arcsec yr^{-1} and distance of $\sim$7 pc in the constellation Monoceros. Its large positive radial velocity implies that just 70,000 years ago, it was at a distance of $\sim$0.25 pc with a proper motion of $\sim$70 arcsec yr^{-1} in the constellation Ursa Major (Mamajek et al. 2015). Our discussion of measuring the parallax is adequate for typical ground-based observations of nearby ultracool dwarfs. For a full astrometric model for high-precision work, see the Gaia model described by Lindegren et al. (2012).

4.2 Astrometry

Astrometry for our purposes is defined as measuring the positions of stars. Over time, the star's proper motion and parallax can be measured, allowing the distance and tangential velocity to be estimated. The Gaia Mission provides highly reliable parallaxes and proper motions for stars with $G \lesssim 20$. As a result of this magnitude limit, the 100 pc Gaia catalog is highly complete for all main-sequence stars down to M7, but most ultracool dwarfs within 100 pc, or even 20 pc, are too faint. We therefore remain highly concerned with measuring parallaxes with targeted observations (Smart 1977; Green 1985; Lindegren et al. 2012; Beichman et al. 2014). From the imagined viewpoint of the solar system barycenter, a star's position on the celestial sphere (right ascension α' and declination δ') is a function of time as the star moves linearly with a proper motion (μ_α, μ_δ):

$$\alpha' = \alpha_0 + \mu_\alpha \times \Delta t / \cos(\delta') \qquad (4.5)$$

$$\delta' = \delta_0 + \mu_\delta \times \Delta t \qquad (4.6)$$

Here, we have made a linear approximation and ignored the radial velocity because we are concerned with short timescales.[3] We typically observe from the Earth (or a near-Earth orbit), so that the parallax effect is important. For a star with parallax p, the observed equatorial coordinates α and δ when the Earth is at positions $X(t)$, $Y(t)$, $Z(t)$:

$$\alpha(t) = \alpha' + p(X(t)\sin\alpha' - Y(t)\cos\alpha')/\cos(\delta') \qquad (4.7)$$

$$\delta(t) = \delta' + p(X(t)\cos\alpha'\sin\delta' + Y(t)\sin\alpha'\sin\delta' - Z(t)\cos\delta') \qquad (4.8)$$

The X, Y, Z positions for the Earth can be calculated with `astropy`; for other positions in the Solar System, such as the Spitzer Space Telescope, the JPL Horizons service can be used. We, therefore, with enough images, can solve for the five parameters α_0, δ_0, μ_α, μ_δ, and p. Figure 4.4 illustrates how the position of W1906 +4011 would vary in time. Theoretically this could be done with just three observations, but in practice with ground-based observations have positional uncertainties of 3–10 mas and many observations are needed to fit the parallax. The proper motion signal builds with time, so with a 10-year baseline and just two measurements, positional uncertainties of $\sim$20 mas can be reduced to $\sim 2\sqrt{2} \approx 3$ mas yr^{-1}. There are three complications to this program. First, we will need to schedule observations to sample the parallax effect, and this means that the most

---

[3] An extraordinary example of how over longer timescales we must also account for the radial velocity is the ultracool dwarf binary WISE J072003.20-084651.2 (Scholz's Star), which currently has a proper motion of $\sim$0.1 arcsec yr^{-1} and distance of $\sim$7 pc in the constellation Monoceros. Its large positive radial velocity implies that just 70,000 years ago, it was at a distance of $\sim$0.25 pc with a proper motion of $\sim$70 arcsec yr^{-1} in the constellation Ursa Major (Mamajek et al. 2015). Our discussion of measuring the parallax is adequate for typical ground-based observations of nearby ultracool dwarfs. For a full astrometric model for high-precision work, see the Gaia model described by Lindegren et al. (2012).

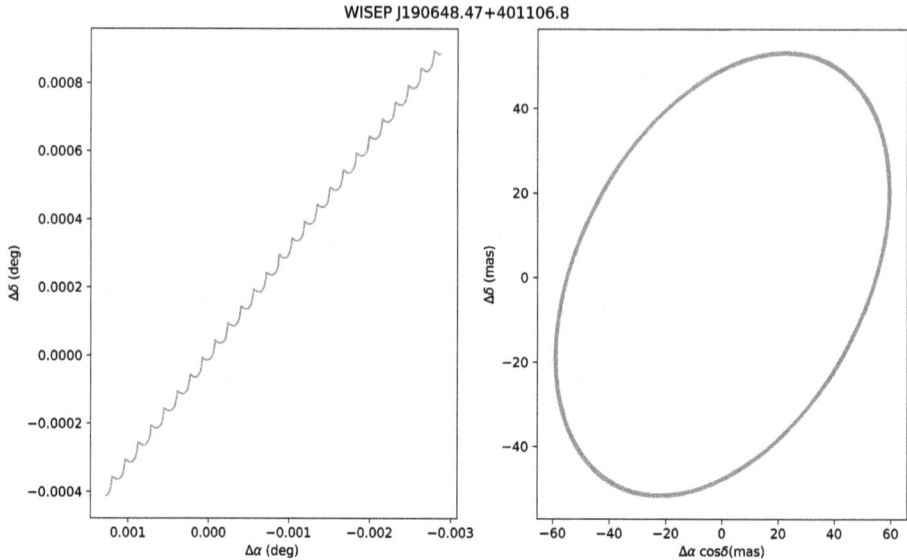

Figure 4.4. The predicted positions for W1906+4011 illustrating the effects of parallax and proper motion. The proper motion signal builds up over time but the parallax signal simply repeats.

important observations need to be scheduled early and late in the night near twilight —if you observe at midnight, when the Earth is at the opposite side of its orbit you would need to observe at noon! Second, we will have to measure our target's position relative to other, more distant stars in the image, but they too have the same parallax effect but at a lower amplitude. If the reference stars are at ∼300 pc, then the correction from relative to absolute parallax is about 0.3 mas. With Gaia, however, the position, proper motion, and parallaxes of the reference stars are known, so that they can be included in the calculation of the reference frame. Finally, a third complication for ground-based observations is differential color refraction. As we have seen, the spectral energy distribution of our target ultracool dwarf will be different from the reference stars, so that the wavelengths observed and the refraction of the target will be different from the reference stars. Our parallax program should include enough observations at different airmasses so that we can solve for this additional parameter, an offset in a known direction.

Finally, we must consider the problem of bias in estimating distances from parallaxes. We measure parallaxes with some uncertainty. Figure 4.5 shows the results of simulations for a sample of stars with parallax measurement errors of 1, 5, and 10 mas: The errors on an individual measurement of the parallax are Gaussian and symmetric.[4] Within 30 pc, the 1 mas sample has $\sigma_p/p < 0.03$ and the estimated distances $1/p$ are well correlated with the true distances with little bias. It is plain,

---

[4] With uncertainties, one could measure a negative parallax even though physically that makes no sense. Negative parallaxes have value in informing us about the measurement uncertainties and should be included in published results.

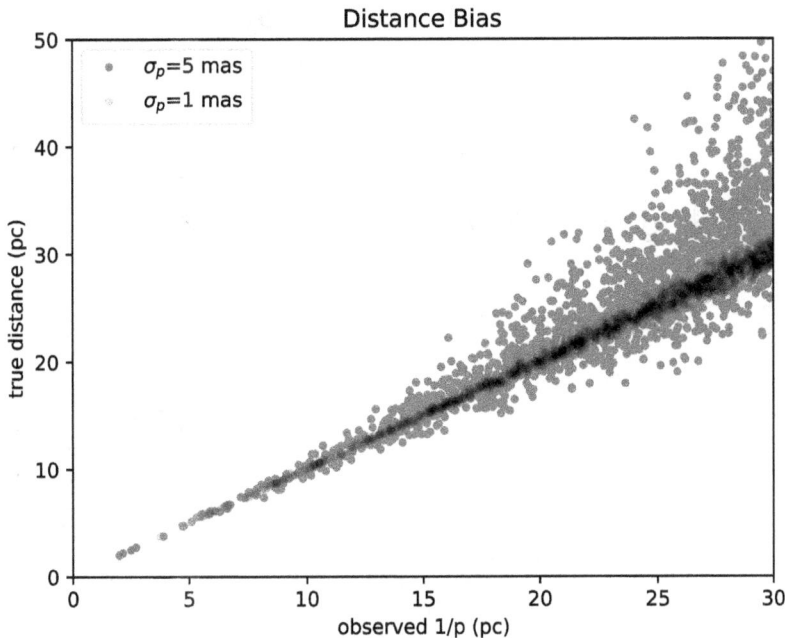

Figure 4.5. An illustration of how the parallax $(1/p)$ is a biased estimator of the true distance. Stars are simulated with a true distance randomly distributed in space but with a parallax measurement error of $\sigma_p = 1$ mas (black), 5 mas (red), or 10 mas (purple). Note how the larger errors mean the many more distant stars are included in the 30 pc sample. For cases where the measured $1/p = 30$ pc but $\sigma_p = 5$ mas, the distances are estimated on average by 10%. This is turn means the absolute magnitudes are biased. For cases where $\sigma_p/p > 0.20$, so many stars have underestimated distances that the bias cannot be well characterized and the parallaxes have very little value.

however, that for the 5 mas simulation, there is considerable bias in the 20 pc $< 1/p <$ 30 pc range, and that many more stars have underestimated distances compared to the number of stars with overestimated distance. The problem is that there are many more distant stars than nearby stars because the volume of a spherical shell scales as d^2. Here, for example, for the stars with 29.0 pc $< 1/p <$ 30.0 pc, $1/p$ underestimates the true distance by $\sim 10\%$ on average. Turning to the sample with 10 mas errors, it is clear that for 20 pc $< 1/p <$ 30 pc, where $\sigma_p/p > 0.2$, the sample is dominated by more distant stars. In this case, the bias is overwhelmingly large, and poorly defined, and the parallaxes have very little value for measuring distances or absolute magnitudes. This effect is known as the Lutz & Kelker (1973) bias and it is discussed in detail by Bailer-Jones (2015) in the context of applying Bayesian priors to measurement of the distances. Actual modeling of samples is necessary.

4.3 Sky Surveys

The odds of discovering a nearby brown dwarf by randomly pointing a telescope is poor: There are 41,254 deg^2 over the entire sky yet there are fewer than 4000 known stars and brown dwarfs within 20 parsecs. Accordingly, sky surveys that

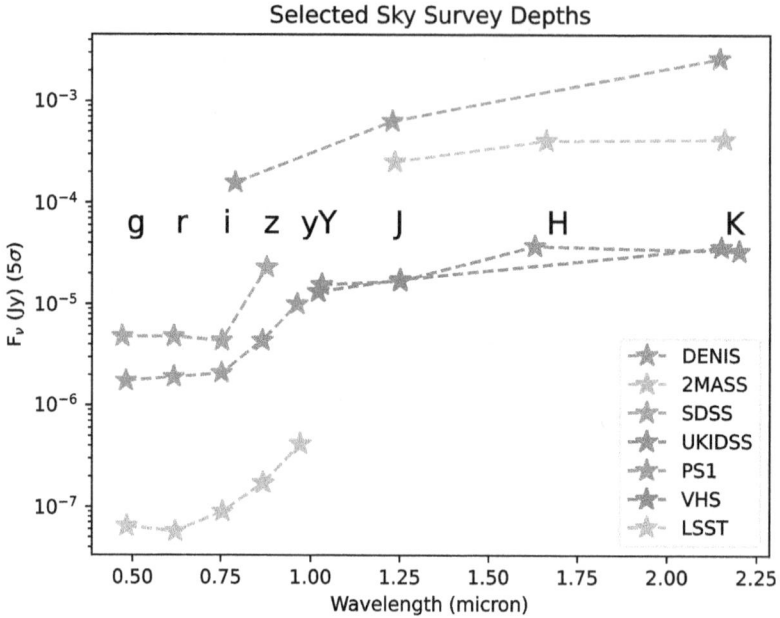

Figure 4.6. The signal-to-noise five detection limits for key optical and near-infrared sky surveys. The PS1 and LSST survey depths are the coadded depths over many years. The WISE and NEOWISE surveys, not shown, reach at least 0.1 mJy at 3.6 μm and 4.5 μm, and as seen in the mid-infrared spectra in Figure 2.3 and 3.14, Y dwarfs are much brighter at 4.5 μm region.

image thousands of square degrees in several filters have been the key tool to discover nearby ultracool dwarfs. Even so, the numbers of nearby objects have to be contrasted with the huge number of distant stars and galaxies: The 2MASS Point Source Catalog includes over 470 million sources and Gaia Data Release 3 includes 1.5 billion sources. Gaia has revolutionized nearby star work because it enables selection directly by measured parallaxes. Most brown dwarfs, however, are too faint to have Gaia parallaxes so we must rely on other methods to pick them out of the huge data sets. The main brown dwarf discovery strategy is to identify the few targets of interest by either their distinct photometric colors or their by large astrometric proper motion when comparing images taken years apart. In either case, once identified, pointed followup can obtain additional photometry, astrometry. and spectroscopy, including parallaxes. Figure 4.6 compare flux densities for 5-sigma sources in selected optical and near-infrared sky surveys as a function of wavelength (filter). To illustrate how these techniques are used in practice, we'll consider two examples, the L1 dwarf WISEP J190648.47 +401106.8 (W1906+4011) and the "archetypal" Y dwarf WISEP J182831.08 +265037.8 (W1828+2650) (Table 4.1).

Figure 4.7 shows 60 arcsec ×60 arcsec images from three major sky surveys centered on the 2010 position of W1906+4011: The optical Sloan Digital Sky Survey (g, r, i), the Two Micron All-Sky Survey (J, H, K_s), and the Wide-field Infrared

Table 4.1. Selected Sky Surveys

Name	Filters	Region	Reference
DENIS	i, J, K	South	Epchtein et al. (1997)
2MASS	J, H, K	All-Sky	Skrutskie et al. (2006)
SDSS	u, g, r, i, z	North	Adelman-McCarthy et al. (2007)
PS1	g, r, i, z, y	North $\delta > -30$	Chambers et al. (2016)
UKIDSS	Y, J, H, K	7500 sq. deg.	Lawrence et al. (2007)
WISE	$W1, W2, W3, W4$	All-Sky	Wright et al. (2010)
VHS	J, K	South	McMahon et al. (2013)
UHS	J, K	North	Dye et al. (2018)
CatWISE	$W1, W2$	All-Sky	Eisenhardt et al. (2020)
unTimely	$W1, W2$	All-Sky	Meisner et al. (2023)

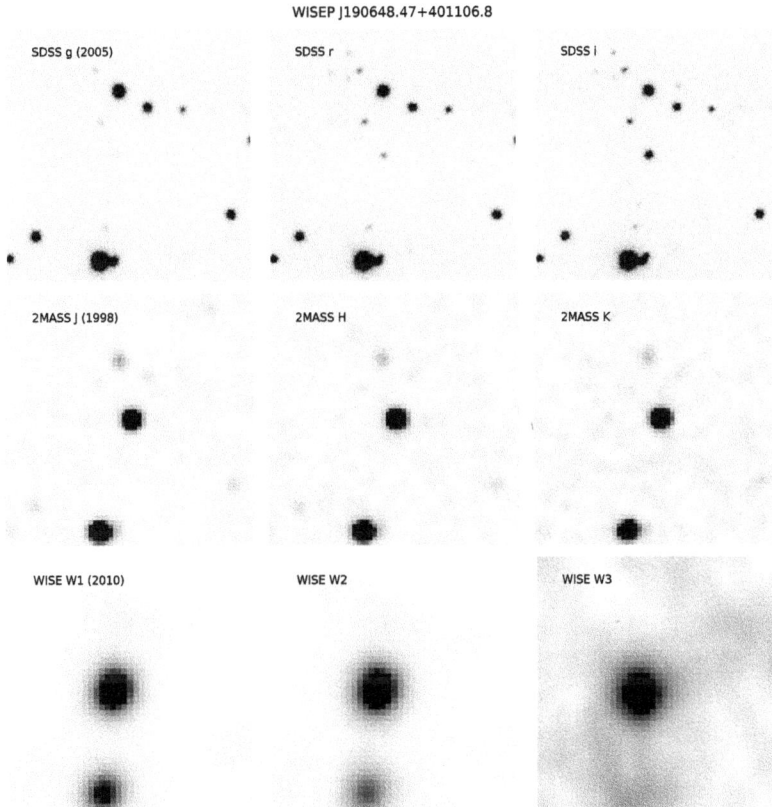

Figure 4.7. Images of the L1 dwarf W1906+40 taken with SDSS (2005 June 8), 2MASS (1998 May 23), and WISE (2010 October 13). It is near the center of each image.

Survey Explorer ($W1$, $W2$, $W3$). Looking at the images, we can see that W1906 +4011 is too faint to be seen in the bluest filter (g), but as we consider the redder optical filters, W1906+4011 is comparable to the background stars in the images. In 2MASS, it is one of the brightest two stars in the images, and in WISE it is the brightest. L dwarfs like W1906+4011 were too faint to be discovered with the 1950s Palomar Sky Survey photographic plates, but they could be efficiently discovered by the red SDSS $i - z$ color, their red 2MASS $J - K$ color, their red PS1 $i - z$, $z - y$ colors, or the combination of WISE with the earlier surveys, bearing in mind that extragalactic sources or reddened Galactic sources could be sources of confusion. Looking carefully at Figure 4.7, we can also see that from 1998 to 2010, W1906 +4011 moved by 6 arcsec. This motion rules out background Galactic or extragalactic sources. Of course, if W1906+4011 had been missed by astronomers analyzing these surveys, it could be selected directly by its Gaia parallax of 59.67 ± 0.10 mas.

Turning to the Y dwarf example, Figure 4.8 shows 120 arcsec ×120 arcsec 2MASS and WISE images for W1828+2650. Now, however, W1828+2650 is seen

WISE J182831.08+265037.7

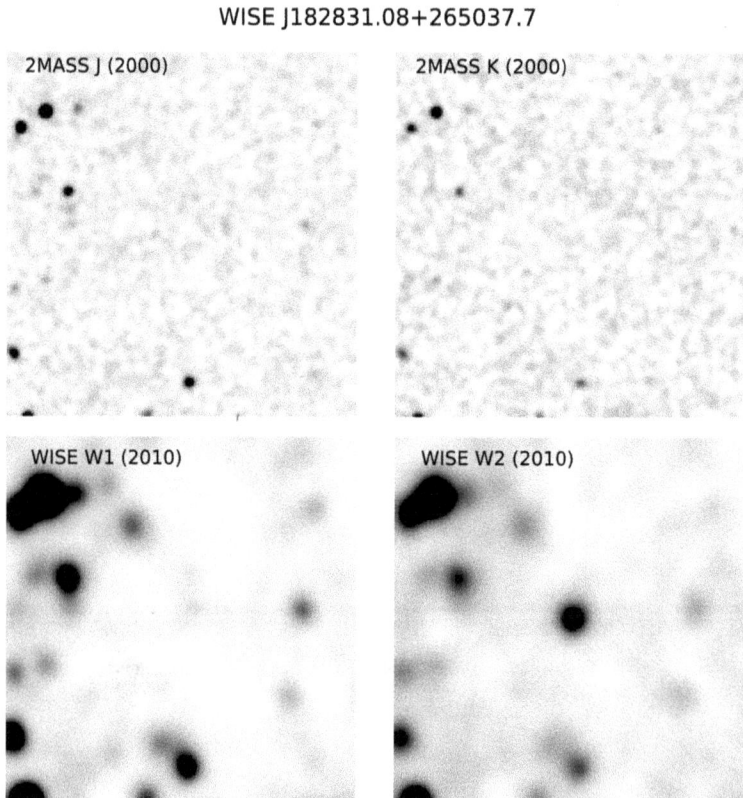

Figure 4.8. Images of the Y0 dwarf W1828+26 taken with 2MASS (2000 May 06), and WISE (2010 March 30). It is the bright source in the WISE W2 image, a faint source in WISE W1, and undetected by 2MASS.

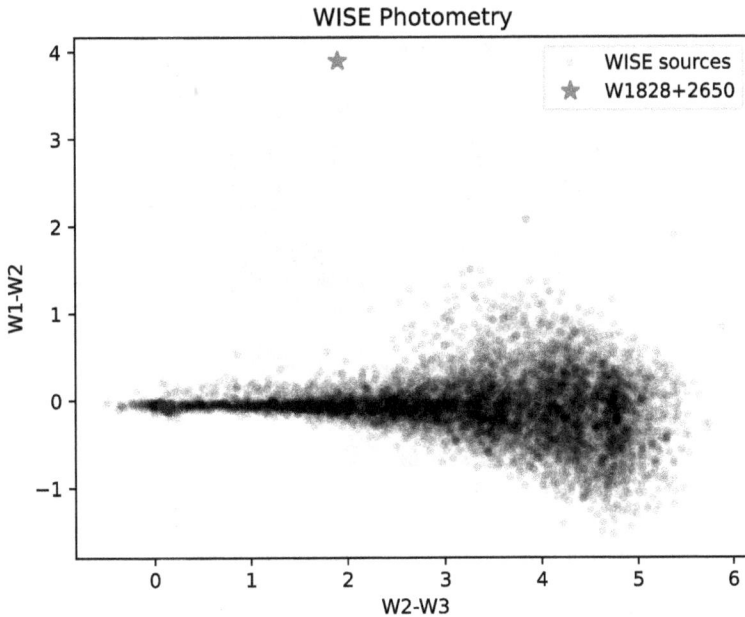

Figure 4.9. WISE photometry of all sources within 20 arcminutes of the Y dwarf W1828+2650. Y dwarfs have distinctive colors that allow them to be identified.

only in the WISE images, because it is less than 1/500 as bright as the detection limit in 2MASS. The WISE filters were designed so that Y dwarfs would look much brighter in the W2 images compared to W1, and indeed this is clearly the case in Figure 4.8. This is quantified in Figure 4.9, where W1828+2650 has much different $W1 - W2, W2 - W3$ colors than other sources in that direction on the sky. As the WISE satellite continued to be operated for several more years as NEOWISE, later W1 and W2 measurements allowed ultracool dwarfs to be selected by their proper motions. They can occur by analysis of a catalog of detected sources with proper motions, such as CatWISE (Eisenhardt et al. 2020), or by visual examinations of the images, such as in the Citizen Science Project *Backyard Worlds: Planet 9* (Kuchner et al. 2017).

Just as parallax-selected samples are biased by the fact that there is more volume and therefore more stars at greater distances, brightness-selected samples will also be biased. This topic is generally called Malmquist bias. It is not very surprising that a survey like 2MASS includes more L dwarfs than Y dwarfs simply because L dwarfs are more luminous in J, H, and K and therefore can be detected in a much larger volume. More subtly, even among say L0 dwarfs, a magnitude-limited sample would overrepresent the more luminous L0 dwarfs and underrepresent the less luminous L0 dwarfs, so that if we calculate the mean absolute magnitude of L0 we will be biased. For ultracool dwarfs, one important effect is that we will tend to include unresolved near-equal luminosity binaries. As with trigonometric parallaxes, this issue should be modeled when interpreting a sample. Butkevich et al.

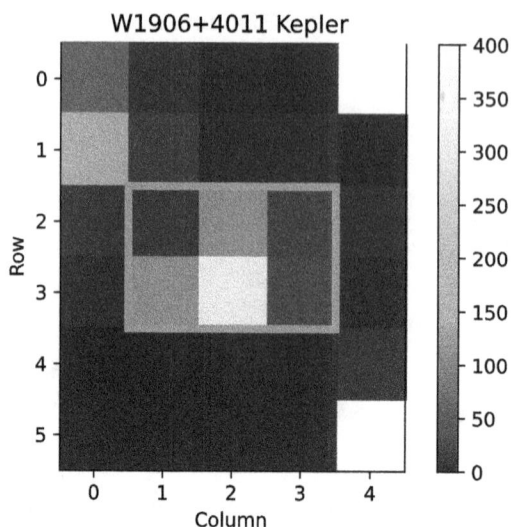

Figure 4.10. Image of W1906+40 as seen by the Kepler Space Telescope. Each pixel is 4 arcsec ×4 arcsec. The background, determined elsewhere on the detectors, has been subtracted off by the pipeline. The central six pixels, outlined in red, are summed to measure the brightness

(2005) gives a detailed discussion of the various forms of Malmquist bias and corrections.

The early sky surveys—DENIS, 2MASS, SDSS—took images of their main footprint on the sky just once. Many later surveys have opened the "time domain" by repeated scans of the sky: NEOWISE, Gaia, PS1, and the future LSST survey, but also optical surveys usually too shallow for brown dwarfs—provide $\sim 10^2 - 10^4$ data points for ultracool dwarfs. Searches for planetary transits, on the other hand, require continuous, high-precision observations. For such work, missions like Kepler and TESS have large fields of view but change pointing rarely. High precision for photometry argues for a strategy in which each target star is kept in the same position on the same pixels as much as possible. The same strategy is often adopted for pointed ground-based observations of planetary transits or searches for brown dwarf variability. In contrast, the sky surveys aim to dither as much as possible in order to ensure that all stars are on the same photometric system. Figure 4.10 shows the Kepler image of W1906+4011: In contrast to the images seen previous, the image is low spatial resolution with 4 arcsec square pixels. Figure 4.11 shows some of the resulting photometry: W1906+4011 is rather faint for Kepler, but time series that were months long were completed. Kepler and TESS obtain data for late-M and L dwarfs in their fields of view as conduct their primary missions of searching for planetary transits of main-sequence stars. These data can be used to measure rotation periods, measure white light flares (Paudel et al. 2018), and search for eclipses and transits (Sagear et al. 2020). For us, a beneficial science result is that searches for transiting Jupiter-mass—or rather, Jupiter radius—exoplanets around stars naturally

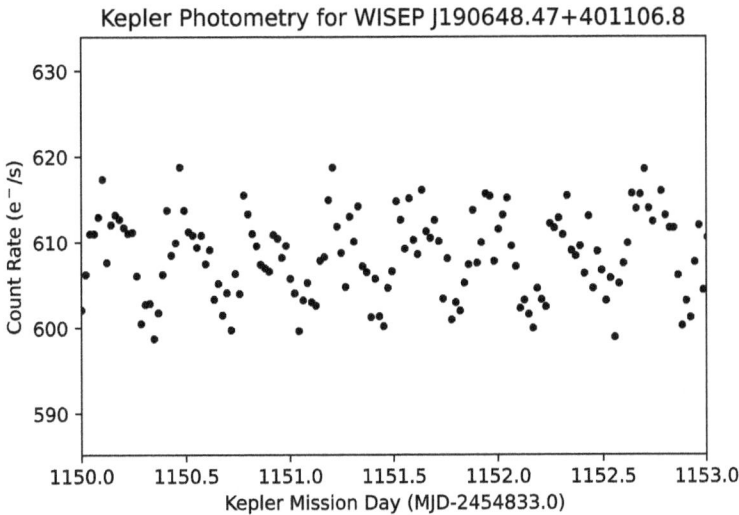

Figure 4.11. Kepler Space Telescope photometry of W1906 + 40, computed from the six pixels labeled in Figure 4.10. Note the periodicity of 0.37 days = 8.9 hr.

identify transiting brown dwarf companions that in turn can be used to test brown dwarf models (Carmichael 2023).

4.4 The Observational H–R Diagram

In Chapter 3, we saw how the bolometric luminosity was correlated with the spectral types. Figure 4.12 shows selected colors as a function of spectral type. Selected absolute magnitudes are shown in Figure 4.13. Cool T and Y dwarfs are very red in $W1 - W2$. Note the remarkable behavior at the L/T transition: Not only do early T dwarfs turn to the blue in $J - K$, but the absolute J magnitude brightens so that the typical T4 dwarf is more luminous at J than an L8-T0 dwarf, even though the bolometric luminosity is (on average) less. HR diagrams were shown for the optical in Figure 1.5 and infrared in Figure 1.6. An example of a color–color diagram is shown in Figure 4.14, where we can see that young brown dwarfs and exoplanets can be much redder than the field ultracool dwarf sequence. This is physically interpreted in terms of clouds but also means that there are important selection effects in sky surveys. Finally, we revisit the spectral type-absolute magnitude relationship in Figure 4.15. We saw in Figure 1.3, that at a given effective temperature, lower mass objects are younger and more luminous. However, when classified with spectral types and measured in the z band, the young M dwarfs are brighter as naively expected but the young L dwarfs are the same or less luminous. The lesson is that spectral type is not synonymous with effective temperature, especially for peculiar dwarfs, and absolute magnitudes in a filter are not synonymous with bolometric luminosities. Each filter will have its own unique behavior.

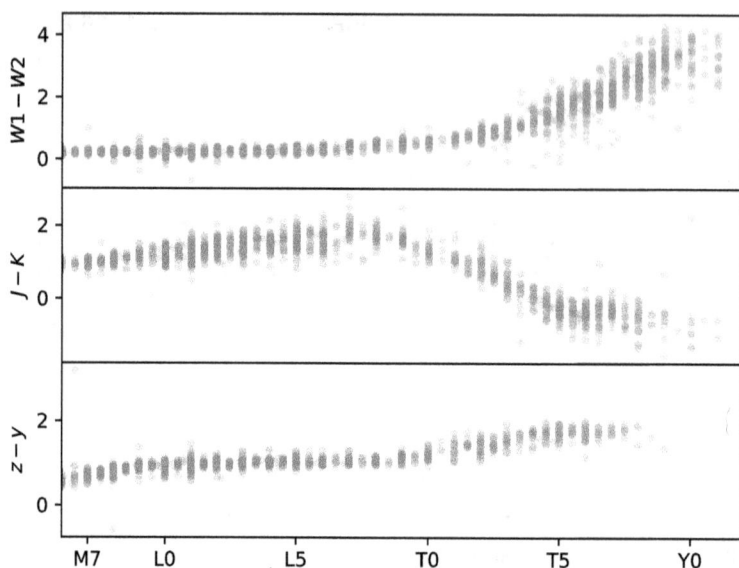

Figure 4.12. Sample colors as a function of spectral type: $z - y$, PS1; $J - K$, MKO; $W1 - W2$, WISE. Only ordinary field dwarfs are shown and known unresolved doubles are excluded.

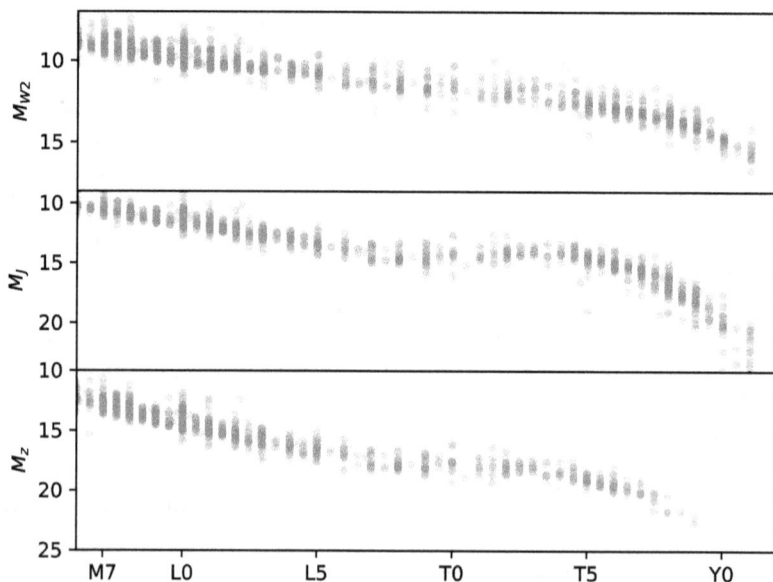

Figure 4.13. Absolute magnitudes as a function of spectral type: M_z, PS1; M_J, MKO; M_{W2}, WISE. Only ordinary field dwarfs are shown and known unresolved doubles are excluded.

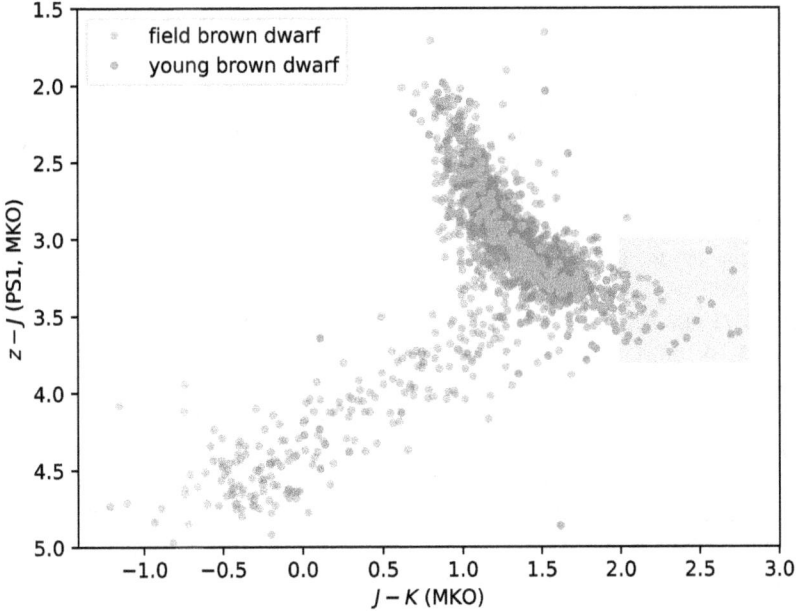

Figure 4.14. Color-color diagram for old field ultracool dwarfs, young ($\lesssim$150 Myr) ultracool dwarfs, and exoplanets. Note the shaded region, where the young sequence runs to much redder $J - K$ colors, presumably due to thick clouds, and that young super-Jupiter exoplanets also have anomalous colors.

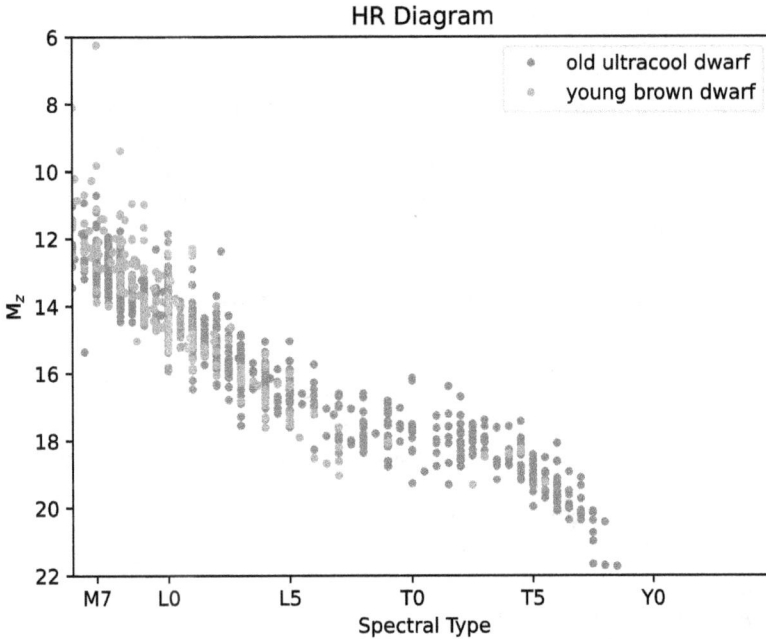

Figure 4.15. Absolute magnitudes M_z as a function of spectral type. Although in the theoretical H–R diagram, young ultracool dwarfs are more luminous at a given T_{eff}, in this observation diagram they are less luminous in M_z. The lesson is that spectral type is not synonymous with effective temperature, especially for peculiar dwarfs, and absolute magnitudes in a filter are not synonymous with bolometric luminosities.

References

Adelman-McCarthy, J. K., Agüeros, M. A., Allam, S. S., et al. 2007, ApJS, 172, 634

Bailer-Jones, C. A. L. 2015, PASP, 127, 994

Beichman, C., Gelino, C. R., Kirkpatrick, J. D., et al. 2014, ApJ, 783, 68

Best, W. M. J., Dupuy, T. J., Liu, M. C., Siverd, R. J., & Zhang, Z. 2020, The UltracoolSheet: Photometry, Astrometry, Spectroscopy, and Multiplicity for 3000+ Ultracool Dwarfs and Imaged Exoplanets, v1.0.0, Zenodo

Bradley, L., Sipőcz, B., Robitaille, T., et al. 2024, astropy/photutils, 1.12.0, 1.12.0, Zenodo, ascl:1609.011

Butkevich, A. G., Berdyugin, A. V., & Teerikorpi, P. 2005, MNRAS, 362, 321

Carmichael, T. W. 2023, MNRAS, 519, 5177

Chambers, K. C., Magnier, E. A., Metcalfe, N., et al. 2016, arXiv e-prints 1612.05560

Cohen, M., Wheaton, W. A., & Megeath, S. T. 2003, AJ, 126, 1090

Dye, S., Lawrence, A., Read, M. A., et al. 2018, MNRAS, 473, 5113

Eisenhardt, P. R. M., Marocco, F., Fowler, J. W., et al. 2020, ApJS, 247, 69

Elias, J. H., Frogel, J. A., Matthews, K., & Neugebauer, G. 1982, AJ, 87, 1029

Epchtein, N., de Batz, B., Capoani, L., et al. 1997, Msngr, 87, 27

Gaia CollaborationMontegriffo, P., Bellazzini, M., et al. 2023, Astronomy and Astrophysics, 674, A33

Green, R. M. 1985, Spherical Astronomy (Cambridge: Cambridge Univ. Press)

Hodgkin, S. T., Irwin, M. J., Hewett, P. C., & Warren, S. J. 2009, MNRAS, 394, 675

Kuchner, M. J., Faherty, J. K., Schneider, A. C., et al. 2017, ApJL, 841, L19

Lawrence, A., Warren, S. J., Almaini, O., et al. 2007, MNRAS, 379, 1599

Lindegren, L., Lammers, U., Hobbs, D., et al. 2012, Astronomy and Astrophysics, 538, A78

Lutz, T. E., & Kelker, D. H. 1973, PASP, 85, 573

Mamajek, E. E., Barenfeld, S. A., Ivanov, V. D., et al. 2015, ApJL, 800, L17

McMahon, R. G., Banerji, M., Gonzalez, E., et al. 2013, Msngr, 154, 35

Meisner, A. M., Caselden, D., Schlafly, E. F., & Kiwy, F. 2023, AJ, 165, 36

Nikolaev, S., Weinberg, M. D., Skrutskie, M. F., et al. 2000, AJ, 120, 3340

Oke, J. B. 1974, ApJS, 27, 21

Paudel, R. R., Gizis, J. E., Mullan, D. J., et al. 2018, ApJ, 858, 55

Persson, S. E., Murphy, D. C., Krzeminski, W., Roth, M., & Rieke, M. J. 1998, AJ, 116, 2475

Sagear, S. A., Skinner, J. N., & Muirhead, P. S. 2020, AJ, 160, 19

Schlafly, E. F., Finkbeiner, D. P., Jurić, M., et al. 2012, ApJ, 756, 158

Simons, D. A., & Tokunaga, A. 2002, PASP, 114, 169

Skrutskie, M. F., Cutri, R. M., Stiening, R., et al. 2006, AJ, 131, 1163

Smart, W. M. 1977, Textbook on Spherical Astronomy (6th eda; Cambridge: Cambridge Univ. Press)

Stephens, D. C., & Leggett, S. K. 2004, PASP, 116, 9

STScI Development Team 2018, synphot: Synthetic photometry using Astropy, 1811, 001 Astrophysics Source Code Library, record ascl:1811.001

Tokunaga, A. T., Simons, D. A., & Vacca, W. D. 2002, PASP, 114, 180

Tonry, J. L., Stubbs, C. W., Lykke, K. R., et al. 2012, ApJ, 750, 99

Wright, E. L., Eisenhardt, P. R. M., Mainzer, A. K., et al. 2010, AJ, 140, 1868

An Introduction to Brown Dwarfs
From very-low-mass stars to super-Jupiters
John Gizis

Chapter 5

Structure and Evolution

5.1 Elements of a Model

The purpose of theoretical models is to predict the luminosity ($\mathcal{L}$), radius ($\mathcal{R}$), surface gravity (g), and other physical quantities as a function of time (t), initial mass ($\mathcal{M}$), and composition (X, Y, Z). Historically important models from the era when brown dwarfs were first observed include Burrows et al. (1993, 1997), Baraffe et al. (1995), Saumon et al. (1996), Chabrier et al. (2000), and newer models are frequently added to the astronomical literature. In this chapter, we'll first review the basic equations of stellar structure, mainly following the book *Stellar Structure and Evolution* (Kippenhahn & Weigert 1990) and then use the open-source code Modules for Experiments in Stellar Astrophysics (MESA) (Paxton et al. 2011, 2013, 2015, 2018, 2019) to compute a family of evolution models using standard physics. We've already seen some of these MESA results in Chapter 1. This capability will allow you to compute quantities that may not be included in journal articles—the central temperature or moment of inertia, for example—and have access to the full interior profile of your model ultracool dwarfs. We will then look at some of the challenges for evolutionary models.

We begin by looking at the standard equations of stellar structure, assuming spherical symmetry so that we have a one-dimensional problem. We define $m(r)$ as the mass enclosed within a radius r and $\rho(r)$ as the mass density as a function of radius. The equation of continuity is

$$\frac{dm}{dr} = 4\pi r^2 \rho \tag{5.1}$$

Integrating from $r = 0$ to $r = \mathcal{R}$ yields the total mass ($\mathcal{M}$):

$$\mathcal{M} = \int_0^{\mathcal{R}} \rho \, 4\pi r^2 dr \tag{5.2}$$

doi:10.1088/2514-3433/ad757ech5

The usual approach in modeling stars, however, is a Lagrangian formulation where we work in terms of the differential mass. Our brown dwarfs will typically have constant total mass ($\mathcal{M}$) throughout their lifetime but the radius will change. The Lagrangian differential equation is

$$\frac{dr}{dm} = \frac{1}{4\pi\rho r^2} \tag{5.3}$$

Next, we consider the brown dwarf to be in hydrostatic equilibrium, so that the pressure gradient balances the gravity. The Lagrangian equivalent of Equation (1.1) is

$$\frac{dP}{dm} = -\frac{Gm}{4\pi r^4} \tag{5.4}$$

The total gravitational energy (E_g) of our object will be

$$E_g = -\int_0^{\mathcal{M}} \frac{Gm}{r} dm \tag{5.5}$$

However, by using Equations (5.3) and (5.4), plus the boundary conditions $r = 0$ at the center and $P = 0$ at the surface, we can also write the gravitational energy without explicitly using the constant G

$$E_g = -\int_0^{\mathcal{M}} \frac{Gm}{r} dm = -3\int_0^{\mathcal{M}} \frac{P}{\rho} dm \tag{5.6}$$

Next, we must consider energy conservation. The mass shell has a net energy per second (power) $L(m)$ entering it from below and a possibly different $L(m) + dL$ leaving it outwards. The source terms are the energy per second deposited into the gas by nuclear reactions (ϵ_n), the loss of energy by any neutrinos created directly by hot plasma (ϵ_ν), and gravitational energy (ϵ_g).

$$\frac{dL}{dm} = \epsilon_n - \epsilon_\nu + \epsilon_g \tag{5.7}$$

The ϵ_ν term can be important in very hot stellar cores but it is zero for ultracool dwarfs. For the nuclear fusion, the situation is simple enough that we can discuss all the significant sources of energy here. As the ultracool dwarf contracts and heats up due to the release, it will first become hot enough to fuse deuterium (^{2}H) with ordinary hydrogen (^{1}H), a reaction that proceeds via the strong nuclear force and has a relatively large cross section:

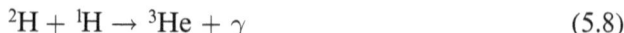

$$^2H + {}^1H \rightarrow {}^3He + \gamma \tag{5.8}$$

The nuclear reaction rates are functions of the composition, density, and temperature of the gas; ϵ_n has the cgs units of erg s^{-1} g^{-1}. Our model, of course, will have to track the corresponding changes in the composition of the gas. The initial abundance of deuterium is $\sim 2 \times 10^{-4}$ so the dwarf will eventually exhaust its supply. Stars on the lower main sequence are powered by the proton–proton (*pp*) chain, fusing hydrogen to helium. The key step is the first step of the proton chain, which requires an improbable tunneling of protons with above-average speeds through a repulsive Coloumb barrier to enable a weak nuclear force interaction:

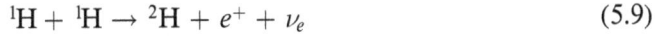

$$^1\text{H} + {}^1\text{H} \rightarrow {}^2\text{H} + e^+ + \nu_e \tag{5.9}$$

The second step of the *pp* chain burns the resulting deuterium quickly

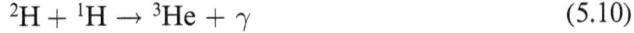

$$^2\text{H} + {}^1\text{H} \rightarrow {}^3\text{He} + \gamma \tag{5.10}$$

In ultracool dwarfs, the central temperature is not high enough for the next steps in the *pp* chain—burning ^{3}He to produce ^{4}He in the *pp*1 chain or ^7Be in the *pp*2 chain. Other nuclear reactions involving the dwarf's initial lithium, beryllium, and boron are possible but these elements are not abundant enough to produce significant energy. Atomic lithium is observable, however, so we need to track this reaction:

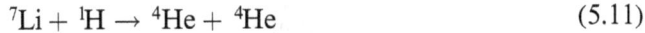

$$^7\text{Li} + {}^1\text{H} \rightarrow {}^4\text{He} + {}^4\text{He} \tag{5.11}$$

The gravitational energy term is a quasi-static process so it can be written as an entropy term:

$$\epsilon_g = T\frac{\partial S}{\partial t} \tag{5.12}$$

The energy is flowing from the hot interior of the dwarf to the cooler outer layers, but we need another equation to calculate the temperature (or thermal) profile. As an example, consider a shell located fairly far from the center of the brown dwarf, so that there are no significant energy sources in the shell and $L(m)$ is constant in this region. The net energy flux L is passing through this shell; the exact same energy flux L is passing through the next shell, and then the same L through a third shell. Even if we somehow know the temperature of the first shell, exactly how much less is the temperature of the second shell or the third shell? To answer this question, we need to know how the energy is being transported. In Eddington's theory of stellar structure, the energy transport is by radiation. In the dense, opaque stellar interior, the individual photons have a short mean free path and travel in a random walk, but on average they diffuse outwards, carrying energy. In this case, we have no need to solve the full, complex equations of radiative transfer for the full distribution of photon frequencies, but instead simply use a diffusion equation:

$$\frac{dT}{dm} = -\frac{3}{16\pi ac}\frac{\kappa_{\text{rad}}L}{r^4T^3} \tag{5.13}$$

where κ_{rad} is the Rosseland mean opacity, and $a = 8m^5k^4/15c^3h^3$ is the radiation constant. Alternatively, the energy transport could be due to conduction, as in dense white dwarfs, with the energy transported by the gas motion collisions between particles. In this case, then the same diffusion equation can be used except a conductive opacity κ_{cond} should be used; the effective total opacity (κ) is $1/\kappa = 1/\kappa_{\text{rad}} + 1/\kappa_{\text{cond}}$.

Instead of thinking in terms of dT/dm, we can think in terms of dT/dP. In hydrostatic equilibrium, pressure is an equally valid measure of position within the dwarf. We have used both m and r as measures of position in the star, but it is also useful to use the pressure P. The actual temperature gradient is defined as

$$\nabla \equiv \frac{d \ln T}{d \ln P} \tag{5.14}$$

and it is related to dT/dm by

$$\frac{dT}{dm} = \frac{dP}{dm}\frac{dT}{dP} = -\frac{GmT}{4\pi r^4 P}\nabla \tag{5.15}$$

The temperature gradient that would exist if all the energy is being transported by radiation is

$$\nabla_{rad} \equiv \left(\frac{d \ln T}{d \ln P}\right)_{rad} = \frac{3}{16\pi acG}\frac{\kappa_{rad} L P}{m T^4} \tag{5.16}$$

We will need to compare this radiative temperature gradient to the adiabatic gradient, where we imagine a parcel of gas changing at constant entropy:

$$\nabla_{ad} \equiv \left(\frac{d \ln T}{d \ln P}\right)_S \tag{5.17}$$

If $\nabla_{rad} > \nabla_{ad}$ the material in the shell is unstable to small perturbations. A "blob" of gas that is hotter than its surroundings will rise and a colder "blob" will sink. This process, called **convection**, transports energy outwards in the star. Our very-low-mass stars will be fully convective for all times, and the brown dwarfs are fully convective during any possible nuclear burning, though they may develop conductive or radiative regions when they become old. On the one hand, this is convenient because it means that we do not need to trace composition changes as a function of position: The burning of ^2H in the central regions reduces its abundance everywhere, and except for edge cases near the D-burning limit, all of the ^2H throughout the dwarf will be burned. The apparent problem is that we seemingly need to develop a complete theory of convection—under conditions that physicists cannot recreate in their laboratories—to calculate the actual temperature gradient ∇. To get a sense of how daunting this prospect is, see the book *An Introduction to Modeling Convection* by Glatzmaier (2013). Most one-dimensional stellar evolution models use a simplified **mixing length theory** (MLT). MLT requires one free parameter, the "mixing length," whose value can be calibrated in solar-type stars using the Sun, but this calibration is not necessarily valid for ultracool dwarfs. Fortunately, in practice this is not a concern for us! The density, and heat capacity, of the material in the interior of ultracool dwarfs is so high that the convective temperature gradient is adiabatic, and so calculating the temperature gradient become a problem in thermodynamics. (For a fully ionized non-relativistic ideal gas, the solution is $\nabla = 2/5 = 0.4$.) Once you have MESA models working, you can verify for yourself numerically that the choice of mixing length parameter does not affect the results.

We have one further problem for determining the temperature profile: None of these treatments of energy transfer are adequate for modeling conditions near the "surface" or "atmosphere" of the ultracool dwarf. As we approach the outer layers from below, $m \rightarrow \mathcal{M}$, $r \rightarrow \mathcal{R}$, $L \rightarrow \mathcal{L}$, but $\rho \rightarrow 0$. As the density drops, the mean free path of the photons becomes longer: The atmosphere often becomes radiative instead of convective, but in any case, the assumption of diffusion, and that we can average over the frequency dependence of opacity breaks down. We must model radiative transfer in more detail both to determine the temperature-pressure profile and to predict the emergent spectrum. Furthermore, convection becomes less efficient—and more difficult to model—so that it is possible that $\nabla \neq \nabla_{\mathrm{ad}}$ in convective regions. This challenge is addressed by modeling the atmosphere separately.

Now that we have defined ∇_{ad}, we can write the gravitational energy source term (Equation (5.12)) in the form that MESA uses (Paxton et al. 2011)

$$\epsilon_g = -TC_P\left[\left(1 - \nabla_{\mathrm{ad}}\chi_T\right)\frac{d \ln T}{dt} - \nabla_{\mathrm{ad}}\,\chi_\rho\frac{d \ln \rho}{dt}\right] \qquad (5.18)$$

where C_P is the specific heat at constant pressure, $\chi_T \equiv (d \ln P / d \ln T)_\rho$, and $\chi_\rho \equiv (d \ln P / d \ln \rho)_T$. You might be surprised that the gravitational constant G does not explicitly appear in a term called the gravitational energy, but remember that the pressure and density of the gas are responding to gravity in hydrostatic equilibrium.

Our goal is to solve these four coupled differential equations. With enough approximations, we could pursue analytic solutions (see Burrows & Liebert 1993 and Auddy et al. 2016), but instead we will now turn to numerical modeling, specifically MESA. Our infinitesimally thin shells become a finite number of thin shells (or "cells"), and the numerical solutions converge to a consistent solution to the four equations and then take steps forward in time (Equation (5.7)) and converge again. Figure 5.1 shows MESA solutions for $r(m)$, $P(m)$, $L(m)$, and $T(m)$ for a $0.05\,\mathcal{M}_\odot$ brown dwarf at two different ages.

Modules for Experiments in Stellar Astrophysics The MESA program (Paxton et al. 2011, 2013, 2015, 2018, 2019) is an open-source Fortran program to solve the equations of stellar structure and evolution. The distribution includes sample calculations of stars, brown dwarfs, and planets. The calculations presented here used version r12778. http://docs.mesastar.org

---

As written, however much of the relevant physics has been hidden in the additional parameters, and it would be impractical for us to re-calculate them all

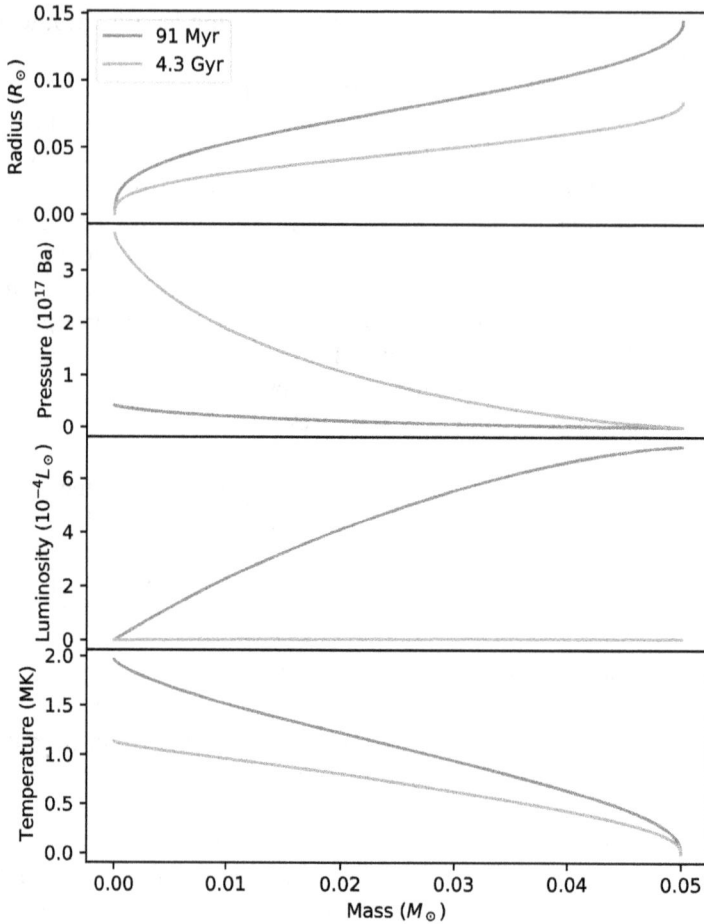

Figure 5.1. MESA models for the interior structure of a $\mathcal{M} = 0.050 \, \mathcal{M}_\odot$ brown dwarf at ages of 91 Myr and 4.3 Gyr. Parameters are shown as a function of the mass coordinate.

from first principles. For example, we need to know statistical and thermodynamic parameters—the pressure (P), specific heat at constant pressure (C_P), the entropy (S), the derivatives (χ_ρ and χ_T), the adiabatic T gradient ∇_{ad}, mean molecular weight (mu), and others—for gas with a known composition (X, Y, Z) as a function of ρ and T—and because we are interested in brown dwarfs, we cannot just assume an ideal gas! Instead, this "equation of state" information has been specially calculated and stored in look-up tables and handled by a MESA subroutine. Similarly, the nuclear reaction networks and the nuclear energy generation rate per unit mass ϵ_n and the opacities (κ) are the stored results of other calculations. In some cases, for some areas of ρ, T parameter spaces, different models have to be combined or even extrapolated. The MESA instrument papers cited above should be carefully examined to get a sense of this. It is worth remembering both that seemingly independent stellar evolution models are often relying on the same underlying opacities, equations of state, as so forth, which may themselves be inaccurate.

One more requirement for solving the differential equations is defining the boundary conditions and the initial conditions (that is, profiles $r(m)$, $P(m)$, $L(m)$, and $T(m)$ at time $t = 0$). The central boundary conditions ($r(m = 0) = 0$, $L(m = 0) = 0$) are straightforward. For the surface, separate atmosphere models —discussed in Chapter 6—that treat the convection and radiative transfer problems in the relatively thin atmosphere are used. These models predict the pressure (P_S) and temperature (T_S) at the base of the atmosphere, and the outermost cell is constrained to be consistent with these "surface" values. The significance of these boundary conditions is discussed extensively later in this chapter. The initial model ($t = 0$) is an extended star of the desired total mass and composition that has a central temperature too low for deuterium burning.

5.2 An Overview of Ultracool Dwarf Models

This section illustrates the evolution of dwarfs with masses in the range $0.001 \, \mathcal{M}_\odot$ through $0.090 \, \mathcal{M}_\odot$ using MESA models. These results are similar to standard published models with cloudless atmospheres, including the COND and Bobcat family of models, and of course are practically the same as the MESA brown dwarf models presented by Paxton et al. (2011). The evolution of the luminosity and the effective temperature for these dwarfs are shown in Figures 1.3 and 5.2.

In the first 10 Myr, a period that is observable in star-forming regions, the very-low-mass stars and more massive brown dwarfs fuse deuterium, as seen Figure 5.4. Higher masses exhaust their deuterium first, as shown in Figure 5.8. After deuterium burning the stars and brown dwarfs contract, similar to the Kelvin–Helmholtz evolution of pre-main sequence stars. For objects just above the D-burning limit, burning is significant in the 10^7–10^8 period and indeed in some cases a lower mass

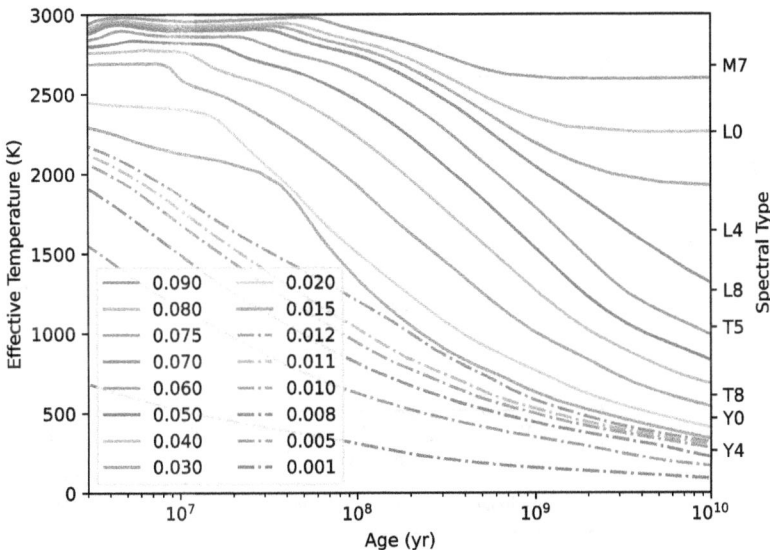

Figure 5.2. The evolution of T_{eff} as a function of time. See Figure 1.3 for luminosity as a function of time.

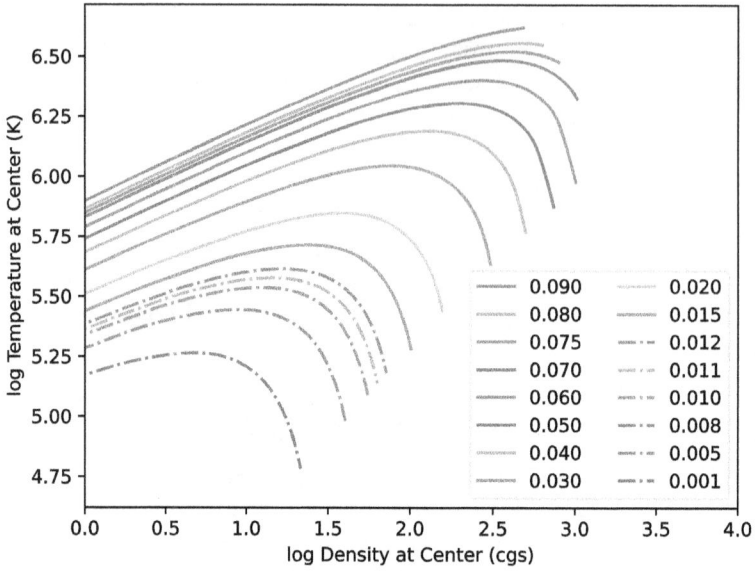

Figure 5.3. Central conditions. The stars and brown dwarfs begin in the lower left at low central density and central temperature and evolve to higher densities and temperatures. Brown dwarfs and planetary mass objects reach a maximum central temperature and then cool. See Figure 1.1 for the central temperature as a function of time.

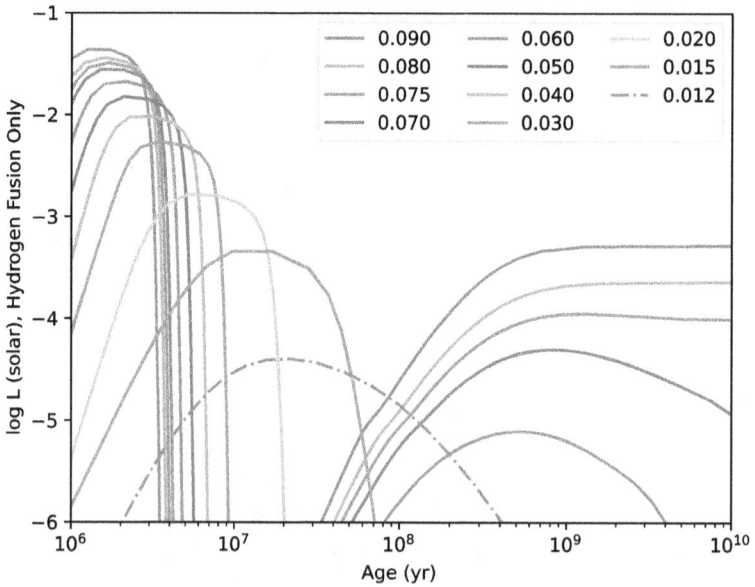

Figure 5.4. The luminosity due to hydrogen-burning: At young ages, this is due to deuterium burning (Equation (5.8)), and at old ages, it is the hydrogen-burning *pp* chain.

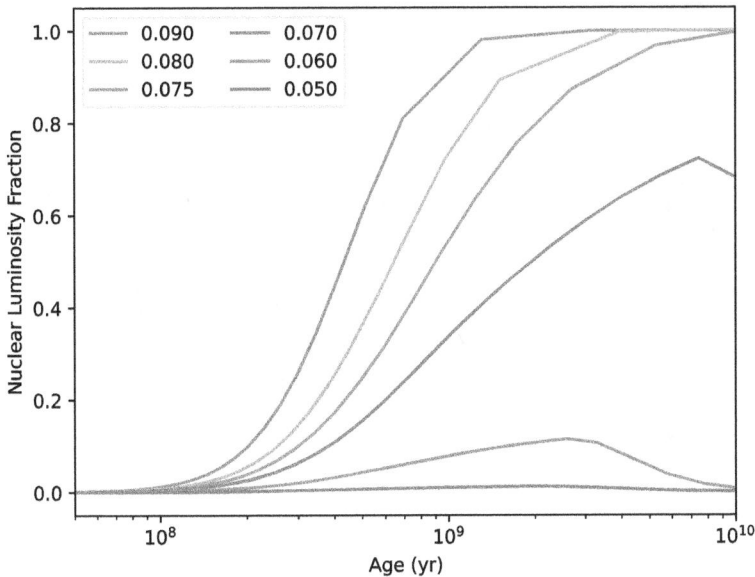

Figure 5.5. The fraction of the total luminosity provided by nuclear fusion. Note that the brown dwarf with $\mathcal{M} = 0.070 \, \mathcal{M}_\odot$ still has 60% of its luminosity from nuclear fusion.

object can be more luminous than a higher one that has exhausted deuterium. The planetary mass objects cool and fade throughout this time period. In this contraction, the central density and central temperature increase until degeneracy and becomes important (Figure 5.3); the stars reach a stable central temperature but the brown dwarfs and planetary mass objects cool (Figure 1.1). Around 10^8 yr, the hydrogen fusion luminosity is increasing for the objects with $\mathcal{M} \gtrsim 0.06 \, \mathcal{M}_\odot$ (Figure 5.4). However, for the $0.060 \, \mathcal{M}_\odot$, this only reaches $\sim 10\%$ of the total luminosity (Figure 5.5). The abundance of lithium (Figure 5.7) and deuterium (Figure 5.8) can be used to distinguish masses around this time. Around 10^9 yr, the very-low-mass stars have reached their main sequence nuclear luminosity. It is worth noting that some nuclear burning is still occurring for brown dwarfs just below the hydrogen-burning limit even at 10^{10} yr. Some high mass brown dwarfs can develop a conductive core. The radius for old objects is shown in Figure 5.6.

5.3 Key Physics: Clouds, Metallicity, and Initial Conditions

5.3.1 The Surface Boundary Condition

The atmosphere plays a key role in regulating the cooling rate of brown dwarfs. From the point of view of constructing models, the atmosphere model provides the pressure–temperature point the interior model must match. The presence of clouds plays a key role in the pressure–temperature structure. Allard et al. (2001) calculated atmosphere models that explored limiting cases of clouds that, even though some of the input opacities are outdated, illustrate the main concerns for evolutionary models and are sufficient for our purposes here. The COND model—the same

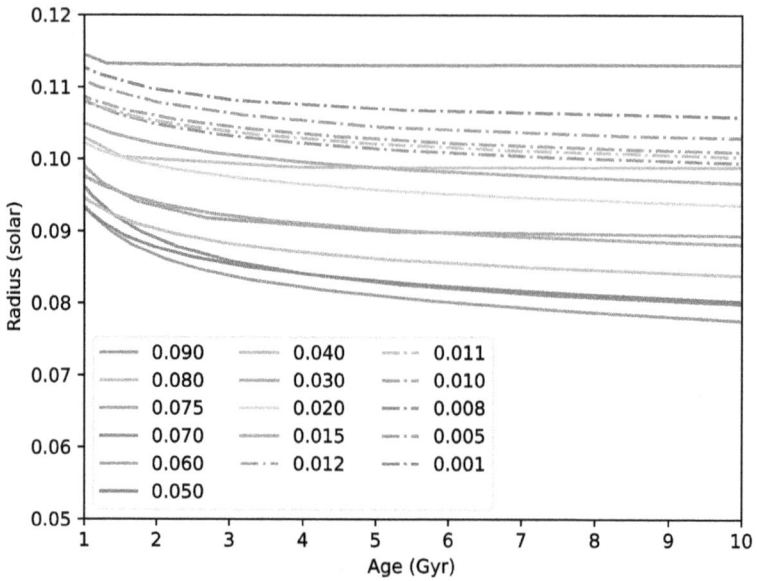

Figure 5.6. The radius for old objects. The minimum radius in this set of models is the 0.060 $\mathcal{M}_\odot$ model.

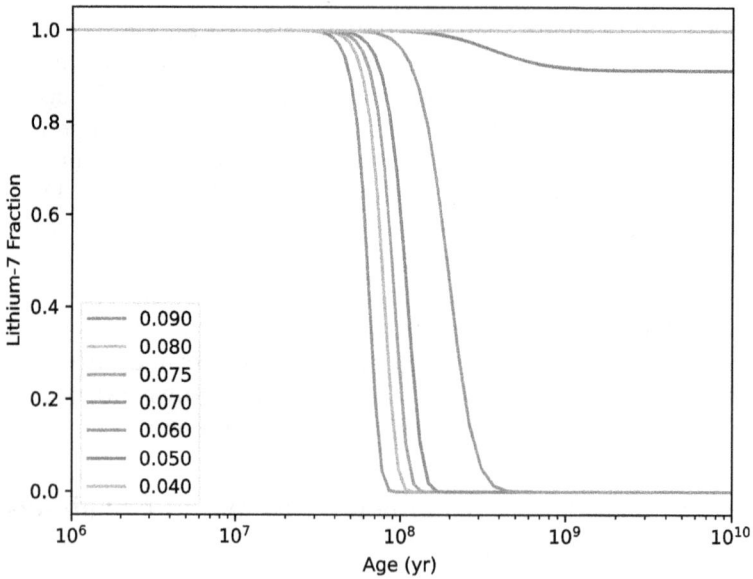

Figure 5.7. The abundance of lithium as a function of time.

boundary conditions we used with MESA—is a limiting case cloudless atmosphere in which condensates form, as predicted by chemistry, but the particles themselves are assumed to sink below the photosphere and contribute no opacity. The DUSTY model was designed to explore the limiting effect of cloud: The particles not only

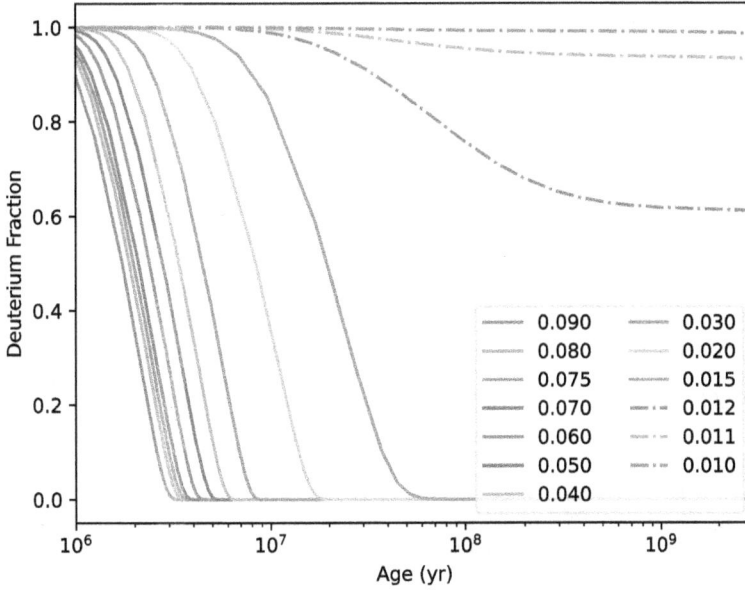

Figure 5.8. The abundance of deuterium as a function of time.

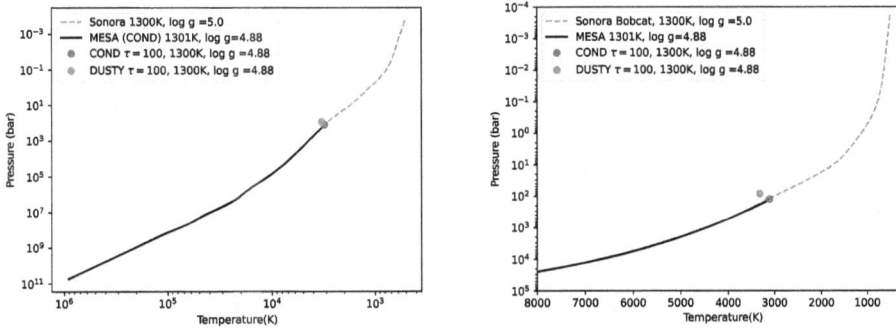

Figure 5.9. An illustration of how the MESA interior solution matches the COND boundary conditions. The DUSTY model would require a profile that is hotter at a given pressure. A Sonora Bobcat model is shown to illustrate how the atmosphere model predicts the profile in the outer regions of the brown dwarf.

condense but remain suspended throughout the atmosphere, adding opacity particularly for cooler and cooler atmospheres. To illustrate this, Figure 5.9 shows the interior pressure–temperature relationship for our 0.030 $\mathcal{M}_\odot$ model at 450 Myr when it has $T_{\text{eff}} = 1300$ K and $\log g = 4.88$. The MESA model has matched the COND boundary condition. We must bear in mind that both the DUSTY and COND atmospheric models are now considered obsolete, but they are sufficient for our purposes here. To guide the eye, we also show a similar, complete Sonora Bobcat atmosphere profile to show how the interior and atmospheric models

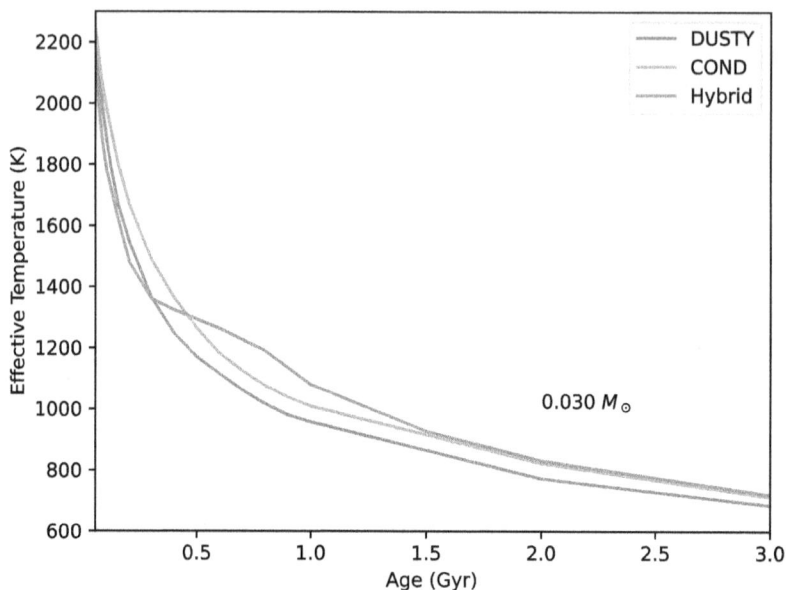

Figure 5.10. Comparison of the extreme cloud (DUSTY) and cloud-free (COND) evolutionary models for 0.030 $M_\odot$. Also shown is a hybrid model that sharply transitions from cloudy to cloud-free at 1300 K. The cloudy reaches lower luminosities and effective temperatures more quickly than the cloud-free model: The hybrid model does as well, but then stalls from 1400 to 1200 K.

together are a complete description of the brown dwarf's profile. We can see, however, that the DUSTY boundary condition for the same T_{eff} and $\log g$ is hotter, so that an interior model that matches the cloudy atmosphere would have to be hotter at all temperatures. This in turn means the evolution of the brown dwarf must be different.

Figure 5.10 shows the effective temperature as a function of time for a 0.030 $M_\odot$ brown dwarf for models with the COND (Baraffe et al. 2003) and DUSTY (Chabrier et al. 2000) boundary conditions. On the one hand, the differences are small in the sense that the T_{eff} differs by only $\sim$10% and the path in the H–R diagram is very similar. The much smaller differences between newer cloudless models with updated opacities and improved physics and the older COND models will have even smaller effects. On the other hand, there is a $\sim$10% effect. These kinds of models also revealed a challenge: While the cloudy atmospheres are needed to explain L dwarf spectra, the T dwarfs are better fit with cloudless atmospheres. Additional challenges with the L–T transition—already revealed in our discussion of the H–R diagram—led to a wide variety of models for the L–T or cloudy-clear transition. Saumon & Marley (2008) presented brown dwarf models where the surface boundary condition is the temperature predicted at $P = 10$ bar their own cloudy or (nearly) clear model atmospheres. Their "hybrid" model, also shown in Figure 5.10, used the cloudy boundary conditions for $T_{\text{eff}} > 1400$ K, the clear boundary conditions for $T_{\text{eff}} \leqslant 1200$ K, and a linear interpolation between them. This relatively sharp transition produces a qualitatively new evolutionary path: The brown dwarf cools rapidly, as in the pure cloudy models, but then stalls at the transition temperature. The brown dwarf has to radiate away its excess interior thermal energy to match the cloudless boundary conditions in order to continue, and this takes additional time compared to non-hybrid models. As we'll see in Chapter 7, brown dwarf counts in the Solar Neighborhood are consistent with this prediction.

Another effect of clouds is on the hydrogen-burning limit. Both the DUSTY and hybrid models report that with cloudy boundary conditions the hydrogen-burning limit is 0.070 $M_\odot$ with $T_{eff} = 1550$ K, lower in both mass and effective temperature than with cloudless boundary conditions. This corresponds to approximately an L4.5 spectral type according to Figure 3.16 in Chapter 3.

5.3.2 Metallicity and the Equation of State

Next, let us consider the effects of metallicity. We know from stars that while many objects are similar in composition to the Sun, some are more metal-rich or metal-poor. Theory predicts that lower metallicity objects are smaller and denser. The consequences of that the hydrogen-burning limit is a higher mass. This is illustrated with the Bobcat Sonora (Marley et al. 2021) models in Figure 5.11. Here, a metal-rich ([Fe/H] = +0.5) 0.072 $M_\odot$ settles onto the main sequence with an effective temperature of 1931K at an age of 10 Gyr. The same mass at solar metallicity is a star at 1699 K but is smaller. The metal-poor ($[Fe/H] = -0.5$) 0.072 $M_\odot$ model is a brown dwarf. Saumon et al. (1994) calculate zero-metallicity

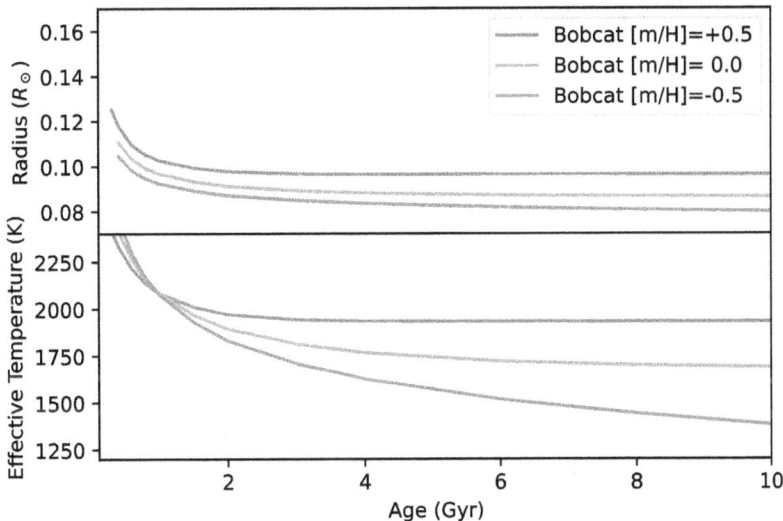

Figure 5.11. Dependence on metallicity from the Sonora Bobcat models. The 0.072 $M_\odot$ model is a brown dwarf at low metallicity but a very-low-mass star at higher metallicities.

object where the hydrogen-burning limit is 0.092 $\mathcal{M}_\odot$. We should also bear in mind the importance of the equation of state. Chabrier & Debras (2021) present a revised equation of state that includes hydrogen–helium interactions. Chabrier et al. (2023) show this leads to denser and cooler structures and a more massive hydrogen-burning limit. The lesson here is that although the hydrogen-burning limit (or hydrogen-burning minimum mass) for the main sequence is a well-defined concept, in practice the theoretical limit is uncertain by a few times 0.001 $\mathcal{M}_\odot$ even for solar metallicity, and it will vary due to metallicity by more than many times 0.001 $\mathcal{M}_\odot$. Furthermore, the difference in model tracks is very small for the first few Gyr. In practice, this means that measuring "the" precise hydrogen-burning limit is a challenge at best.

5.3.3 Initial Conditions

The STAR-LIKE research paradigm points us to initial conditions that correspond to the fragmentation and collapse of a molecular cloud core. What are the initial conditions of such an object? A full answer must depend on the details of the accretion of the mass over time. The evolutionary models shown so far begin with a large, convective configuration that is warm but not hot enough for deuterium fusion. Even here, there is uncertainty: What radius should be chosen? How much entropy per particle? Baraffe et al. (2002) explored a range of initial conditions, resulting in significant differences for the first 10^6 yr, but the evolutionary tracks agreed after a few million years. Baraffe et al. (2009) presented models in which mass is accreted episodically over the first $\sim 10^4$ yr —that is a few bursts of high accretion rate rather than a continuous but lower accretion rate—that produce a spread in the H–R diagram at 1 Myr. The message for us is that we must be cautious in interpreting evolutionary models at ages <10 Myr.

The SUPER-JUPITER research paradigm suggests that we should be searching for signatures of giant planet formation. Marley et al. (2007) discuss the consequences of core accretion scenarios for the initial conditions. In this planet scenario, the accreted gas is thought to radiate away its gravitational potential energy before reaching the surface of the planet. The result is a much lower entropy that the STAR-LIKE initial conditions, or more simply, a cold star compared to the hot-start star-like models. Remarkably, this effect is most important for higher mass planets: "Jupiter mass planets (1 $\mathcal{M}_{Jup}$) align with the conventional model luminosity in as little at 20 million years, but 10 $\mathcal{M}_{Jup}$ planets can take up to 1 billion years to match commonly cited luminosities." In Figure 5.12, we show models calculated by Spiegel & Burrows (2012). Here, we can see that in this particular set of models the hot-start 10 $\mathcal{M}_{Jup}$ model is about 20 times more luminous than the cold-start one even at 100 Myr. Here, we should note that the known directly imaged planets do not seem to be anything like this subluminous—they likely result from hot, or at least warm, starts. This may be a selection bias (Fortney et al. 2008).

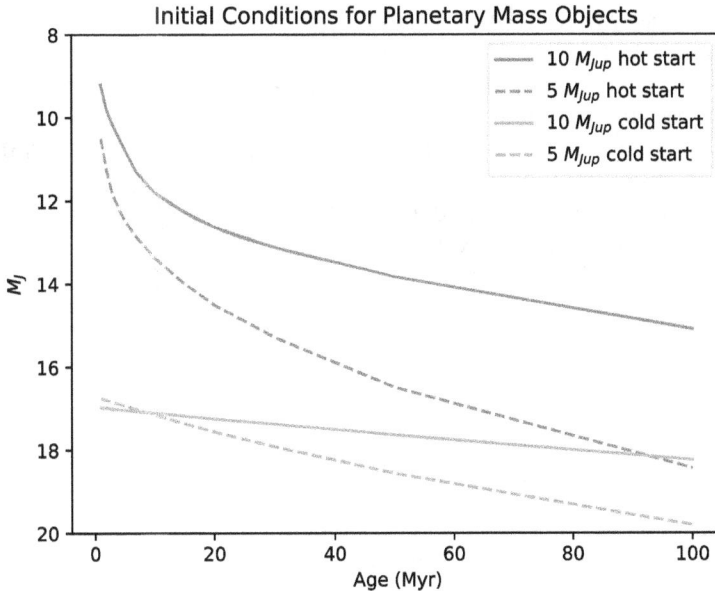

Figure 5.12. Evolutionary models from Spiegel & Burrows (2012) for $10\tilde{M}$jup and 5 $\mathcal{M}_{Jup}$ objects with high-entropy "hot" initial conditions, corresponding to a star-like formation. and low-entropy "cold" initial conditions, corresponding to (some) planet formation models. The absolute J magnitude is shown, which also requires bolometric corrections from model atmosphere synthetic spectra. The models shown use cloud-free atmospheres and are solar metallicity.

References

Allard, F., Hauschildt, P. H., Alexander, D. R., Tamanai, A., & Schweitzer, A. 2001, ApJ, 556, 357

Auddy, S., Basu, S., & Valluri, S. R. 2016, AdAst, 2016, 574327

Baraffe, I., Chabrier, G., Allard, F., & Hauschildt, P. H. 1995, ApJL, 446, L35

Baraffe, I., Chabrier, G., Allard, F., & Hauschildt, P. H. 2002, A&A, 382, 563

Baraffe, I., Chabrier, G., Barman, T. S., Allard, F., & Hauschildt, P. H. 2003, A&A, 402, 701

Baraffe, I., Chabrier, G., & Gallardo, J. 2009, ApJL, 702, L27

Burrows, A., Hubbard, W. B., Saumon, D., & Lunine, J. I. 1993, ApJ, 406, 158

Burrows, A., & Liebert, J. 1993, RvMP, 65, 301

Burrows, A., Marley, M., Hubbard, W. B., et al. 1997, ApJ, 491, 856

Chabrier, G., Baraffe, I., Allard, F., & Hauschildt, P. 2000, ApJ, 542, 464

Chabrier, G., Baraffe, I., Phillips, M., & Debras, F. 2023, A&A, 671, A119

Chabrier, G., & Debras, F. 2021, ApJ, 917, 4

Fortney, J. J., Marley, M. S., Saumon, D., & Lodders, K. 2008, ApJ, 683, 1104

Glatzmaier, G. A. 2013, Introduction to Modelling Convection in Planets and Stars (Princeton, NJ: Princeton Univ. Press)

Kippenhahn, R., & Weigert, A. 1990, Stellar Structure and Evolution (Berlin: Springer)

Marley, M. S., Fortney, J. J., Hubickyj, O., Bodenheimer, P., & Lissauer, J. J. 2007, ApJ, 655, 541

Marley, M. S., Saumon, D., Visscher, C., et al. 2021, ApJ, 920, 85

Paxton, B., Bildsten, L., Dotter, A., et al. 2011, ApJS, 192, 3

Paxton, B., Cantiello, M., Arras, P., et al. 2013, ApJS, 208, 4

Paxton, B., Marchant, P., Schwab, J., et al. 2015, ApJS, 220, 15

Paxton, B., Schwab, J., Bauer, E. B., et al. 2018, ApJS, 234, 34

Paxton, B., Smolec, R., Schwab, J., et al. 2019, ApJS, 243, 10

Saumon, D., Bergeron, P., Lunine, J. I., Hubbard, W. B., & Burrows, A. 1994, ApJ, 424, 333

Saumon, D., Hubbard, W. B., Burrows, A., et al. 1996, ApJ, 460, 993

Saumon, D., & Marley, M. S. 2008, ApJ, 689, 1327

Spiegel, D. S., & Burrows, A. 2012, ApJ, 745, 174

An Introduction to Brown Dwarfs
From very-low-mass stars to super-Jupiters
John Gizis

Chapter 6

Atmospheres

6.1 The Problem

We have seen the rich spectra of M, L, T, and Y dwarfs in Chapter 3. Our problem is to understand how this light is emitted by the brown dwarf and what it can tell us about the brown dwarf itself. The modeling methods have similarities to those used for stars and planets, and can be explored in far more detail but different contexts in textbooks such as *Theory of Stellar Atmospheres* (Hubeny & Mihalas 2014) and *Exoplanetary Atmospheres* (Heng 2017).

Figure 6.1 shows a cartoon sketch of the atmosphere. The atmosphere is heated from below by the hot interior of the brown dwarf. The atmosphere processes this energy flux and emits it as photons into space with some spectral energy distribution $\mathcal{F}_\lambda$. The atmosphere is very thin compared to the size of the brown dwarf, so we can treat this as a one-dimensional problem with spatial coordinate z originating ($z = 0$) at whatever we choose to be the base of the atmosphere, and consider that we are modeling some representative cross-section, such a unit area. The atmosphere is in a steady-state equilibrium, so that the energy flux from the deep interior $\mathcal{F}_{\text{int}}$ equals the emitted radiation $\mathcal{F}$.

Each layer of the atmosphere is in hydrostatic equilibrium: The downward force of gravity is balanced by the pressure gradient. Taking the atmosphere to be an ideal gas:

$$\frac{dP}{dz} = -g\rho = -g\frac{\mu P}{kT} \tag{6.1}$$

In a brown dwarf, both pressure (P) and temperature (T) will be a function of height (z). For some insight, let's consider an isothermal atmosphere, so that T is a constant. Integrating Equation (6.1):

$$P(z) = P_0 e^{-g\mu z/kt} = P_0 e^{z/H} \tag{6.2}$$

doi:10.1088/2514-3433/ad757ech6

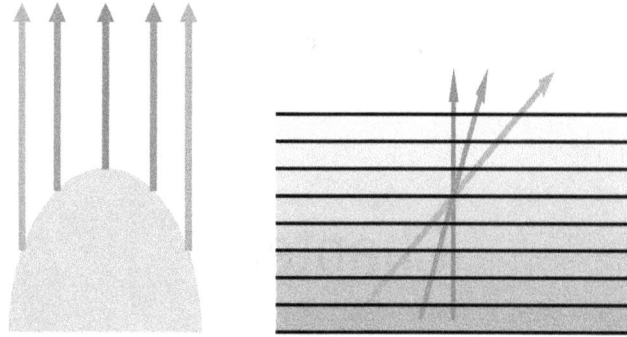

Figure 6.1. We wish to calculate the flux emitted by a spherical brown dwarf as observed by a distant observer (left). We simply the problem to the one-dimensional plane parallel atmosphere with a finite number of layers (right). All outgoing beams ($\mu > 0$) contribute to the observed spectrum.

We see that the pressure profile is exponential with a scale height $H = kT/g\mu$. For a value of $T = 1000$ K, $\log g = 5$, and $\mu = 2.362 m_{\rm H}$, this is 3.5 km or $5 \times 10^{-5} \mathcal{R}_{\rm Jup}$. An actual brown dwarf atmosphere will not be isothermal, but the temperature varies slowly enough that an exponential is good approximation locally. Most importantly, as we discussed in the previous chapter, the pressure monotonically decreases outwards so that P is a unique description of a layer in the atmosphere.

6.2 Radiative Transfer

Imagine for the moment that we know—or guess—the pressure–temperature (PT) profile—that is, we know P and T for each layer. Let us also imagine we also know the composition of the atmosphere—not merely the relative abundance of the constituent elements, as in Section 2.3—but the relative numbers of all the atoms and molecules (H_2, He, CO, CO_2, CH_4, H_2O, etc.) and therefore corresponding the mean molecular weight. (We also need to know all the abundance and sizes of any aerosols—small condensate particles that make up clouds—that are present, but for the moment let us assume this atmosphere has none.) In this case, the ideal gas law tells us the mass density ρ, and if we furthermore know g, we then have z for each layer of the atmosphere from hydrostatic equilibrium (Equation (6.1)). We have now simplified our problem to be that of calculating the radiation emitted by these horizontal slabs of gas—a one-dimensional radiative transfer problem, as seen in Figure 6.1.

The intensity of the radiation field at any point z (or pressure P) is I_λ, with units erg s^{-1} cm^{-2} μm^{-1} ster^{-1} if we work in terms of wavelength. Instead of angle θ, with $\theta = 0$ perpendicular to the atmosphere, it is convenient to work in terms of $\mu = \cos\theta$. We must also consider the optical depth τ:

$$\tau \equiv \int \kappa \, dm = \int \frac{\kappa\rho}{\mu} dz \qquad (6.3)$$

Note that τ decreases outwards and that it is a function of wavelength. The photons that our telescopes observe typically arise at $\tau \approx 1$; in brown dwarfs, τ is a strong function of wavelength and we will therefore observe photons from many different layers of our model atmosphere. The time-independent solution for the radiation along a beam in direction μ is:

$$\mu \frac{dI}{d\tau} = I - S \qquad (6.4)$$

Note that if there were no sources of new radiation in the layer, then the solution would simply be $I(\tau) \propto e^{-\tau}$; the opacity of each layer reduces the intensity. The source function S represents the added intensity due to emission by the gas in the layer. Although we have greatly simplified the problem, we still have a tremendous amount of physics swept into Equation (6.4). For example, the spectra in Chapter 3 have already convinced us that the opacity due to molecules must be a strong function of wavelength, but those spectra were relatively low resolution. In reality, each molecule may have $\sim 10^6 - 10^8$ discrete lines, each of which is pressure broadened due to the presence of other molecules. All radiative transfer and model atmosphere codes must make approximations when trying to solve this problem. Open-source codes such as petitRADTRANS (Mollière et al. 2019, 2020) and picaso (Batalha et al. 2019; Mukherjee et al. 2023) have extensive histories for brown dwarfs and directly imaged exoplanets. We must emphasize that both the opacity due to molecules and the PT profile are crucial ingredients: As an extreme example, Figure 6.2 shows that an isothermal PT profile will produce a featureless blackbody spectrum even if there are many molecules present.

Radiative Transfer The Python package picaso (Batalha et al. 2019; Mukherjee et al. 2023) available at https://github.com/natashabatalha/picaso is used for many of the calculations in this chapter. petitRADTRANS (Mollière et al. 2019, 2020) is an alternative Python package available at https://petitradtrans.readthedocs.io/. Both packages include extensive tutorials and documentation.

We now have the ability to turn a model atmosphere—that is, a PT profile and the molecular abundances at each pressure—into an emergent spectrum which we can compare to models. Two strategies now present themselves. The first is to use all our knowledge of physics and chemistry to predict the PT profile and abundances. In stellar astrophysics, this is called a "model atmosphere" and in planetary science a "climate model." That is, we might assume, for example, that $T_{\rm eff} = 1000$ K, $\log g = 5.0$, and a solar composition, meaning elemental abundances according to one of sources in Table 2.1. Just in the interiors problem, to get the temperature, we will need to model energy transport by radiation, check if the layer is convectively unstable, and if so apply a model of convective energy transport. We will also need a chemical model to calculate the molecular abundances at each layer given its pressure and temperature, and look up

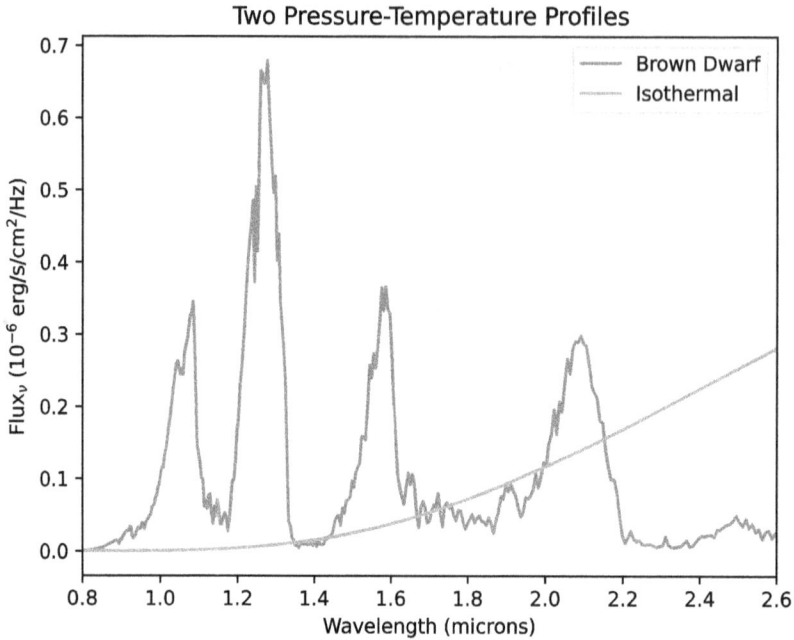

Figure 6.2. picaso calculated emission spectrum for a Sonora Bobcat brown dwarf PT profile with equilibrium chemistry ($T_{\mathrm{eff}} = 1000$ K) compared to an isothermal (1000 K) profile with the same molecular abundances in each layer. While the presence of H_2O and CH_4 leads to a complex T dwarf-like spectrum in the more realistic Bobcat PT profile, there are no absorption features in the isothermal profile.

the cross-section and opacity of each atom and molecule (Figure 6.3). The chemical models will predict the formation of solid and liquid particles, and so we will also need a model for what happens to these "grains" or "dust"—Do they sink deep into the atmosphere or are they suspended in clouds, and if, at what pressure layers? What size are these grains and what are their opacity? Just as the interior model required a numerical strategy to converge on a solution, so too does this atmosphere model. picaso includes a climate model package. One conceptual strategy, in regions where convection does not occur, is to check the time-dependent radiative heating:

$$\frac{\partial T}{\partial t} = -\frac{1}{\rho C_P} \frac{\partial \mathcal{F}_{\mathrm{net}}}{\partial z} \tag{6.5}$$

The Sonora Bobcat profile is in radiative–convective equilibrium ($dT/dt = 0$) but an arbitrary profile, like the isothermal, that we guess would not be. In an actual physics situation, we would expect different layers to be heated or cooled until the entire atmosphere reached a time-independent equilibrium. T_{eff} must be chosen so that we know how much energy (erg s^{-1} cm^{-2}) is entering the bottom of the atmosphere—and correspondingly being radiated into space. We might then repeat the calculation with the same composition with other values for parameters T_{eff} and

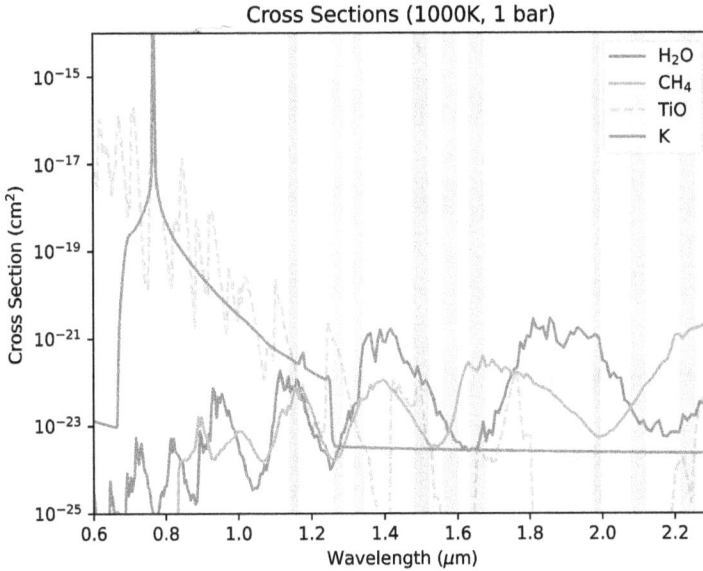

Figure 6.3. Cross sections per particle for H_2O, CH_4, K, and TiO at $T = 1000$ K and $P = 1$ bar. Actual spectra are shaped by contributions at many temperatures and pressures, but these can be compared to the observed T dwarf spectra in Chapter 3. The indices used in the Burgasser et al. (2006) T dwarf classification system are shown are vertical shaded bands. Note that K is an important opacity source for the red optical because it is pressure-broadened over a wide range of wavelengths. See Allard et al. (2024) for recent results on K broadening. TiO bands would be important if TiO were present in the upper atmosphere.

$\log g$, and perhaps even consider different compositions, such as scaling the solar abundances up or down or using a different carbon-to-oxygen ratio, and perhaps vary other parameters that enter into our model as discussed later in the chapter. This has been called the "grid" model approach in the brown dwarf community, because we will end up computing models to discretely sample a large grid in parameter space. We'll discuss some current grid models in Section 6.5. Of course, there is no guarantee that our approach will match the actual observed spectra perfectly: There may be shortcomings in the underlying theoretical or laboratory data, some aspect of our models (such as radiative transfer approximations) might be too simplistic, perhaps our input assumptions were wrong (solar composition), or perhaps we are missing some physical effect (clouds? magnetic fields? mixing? rotation? something unknown?) entirely. In any case, given a model grid, we can search for the "best fit" synthetic spectra for any given observed spectrum and thereby infer the best fit parameters for that brown dwarf.

The alternative strategy to this grid search approach is to search for a $\log g$ value, PT profile and an abundance profile that matches the observed spectrum. This general strategy is called a "retrieval" analysis in the brown dwarf field and is popular in the exoplanet community as well. It has the advantage of being driven by the data and does not assume we can compute the PT profile or abundances from first principles. Even if our knowledge of brown dwarf atmospheres was perfect, in the STAR-LIKE paradigm, we expect a range of elemental abundances that we would like to measure, and we would like to investigate the SUPER-JUPITER metal-enriched scenarios as well. If we can measure the major molecules that include C and O, then we can get the C/O ratio from a retrieval analysis without assuming one beforehand. Because our models are still developing, the retrieval analysis has the potential for identifying shortcomings in our modeling of brown dwarf atmospheres with this approach, whether chemical or energy-transfer related. petitRADTRANS includes a retrieval package and a tutorial to apply it to an exoplanet's thermal emission spectrum.

6.3 Key Physics

6.3.1 Equilibrium Chemistry

A good starting point for understanding what molecules exist is to calculate abundances of molecules given the solar elemental abundances of elements (Table 2.1) in each layer of our model atmosphere. Given the elemental abundances, and taking the pressure and temperature, the network of reactions must be solved to give the abundance of each atom, ion, molecule, and free electrons. The reactions are assumed to be in chemical equilibrium—more specifically, they are in *thermochemical* equilibrium to distinguish from *photochemical* equilibrium, which may be more appropriate the upper layers of in irradiated planets. The equilibrium results can be calculated beforehand: picaso, for example, is distributed with pre-computed tables as a function of temperature and pressure. By far the most abundant molecule will be H_2:

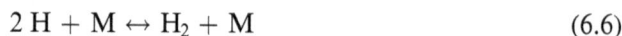

$$2\,H + M \leftrightarrow H_2 + M \tag{6.6}$$

Here, there is a third body involved in the reaction (M). Molecular hydrogen, however, has little opacity though collision-induced absorption (CIA) is included in models and may be important in metal-poor and high-density atmospheres. We know that M dwarfs and L dwarfs show CO features and that T dwarfs have strong CH_4 features, so the reactions forming carbon monoxide and methane are of special interest:

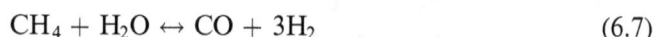

$$CH_4 + H_2O \leftrightarrow CO + 3H_2 \tag{6.7}$$

Figure 6.4 shows the results for solar composition (Lodders 2010) at three different pressures. A complete accounting of carbon includes other molecules such as CO_2 but typically they are much less abundant in brown dwarfs. Model atmospheres in chemical equilibrium mean that each layer of the atmophsere is in the chemical equilibrium approriate for its pressure and temperature. Table 6.1 gives references for the major chemical reactions for selected elements.

Opacities ExoMol is a database of molecular line lists and related modeling software (Tennyson & Yurchenko 2017) major molecules. These are necessary for brown dwarf model atmospheres. In this book, we use the molecular cross-sections and opacities as packaged at lower resolution by picaso. https://www.exomol.com

---

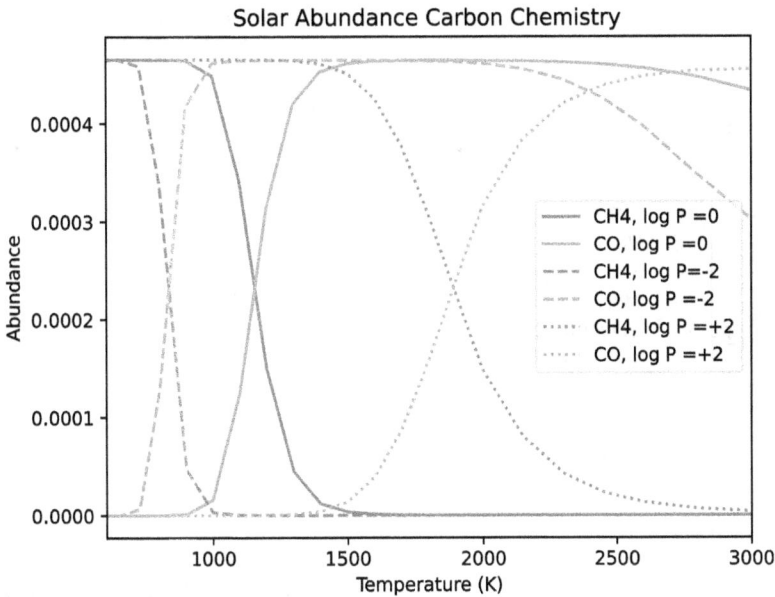

Figure 6.4. An illustration of chemical equilibrium showing CO and CH$_4$ as a function of temperature at three different pressures.

Table 6.1. Elements of Interest

Element	Chemistry	Reference
Li	Li,LiH, LiF, LiOH, LiCl	Gharib-Nezhad et al. (2021)
C	CO, CH$_4$, CO$_2$	Lodders & Fegley (2002)
N	NH$_3$	Lodders & Fegley (2002)
O	H$_2$O, CO, clouds	Lodders & Fegley (2002)
Mg	Mg, MgH, clouds	Visscher et al. (2010)
Si	Si, SiO, clouds	Visscher et al. (2010)
K	K	Burrows & Sharp (1999)
Ti	Ti, TiO	Lodders (2002)
V	VO	Lodders (2002)
Fe	Fe, FeH	Visscher et al. (2010)

6.3.2 Condensate Clouds and Rainout Chemistry

The chemical calculations already discussed predict the formation of solid and liquid particles (Figure 6.5). Iron, for example, may condense out of the gas phase:

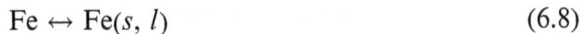

$$Fe \leftrightarrow Fe(s, l) \tag{6.8}$$

Similarly, magnesium, oxygen, and silicon can form silicates:

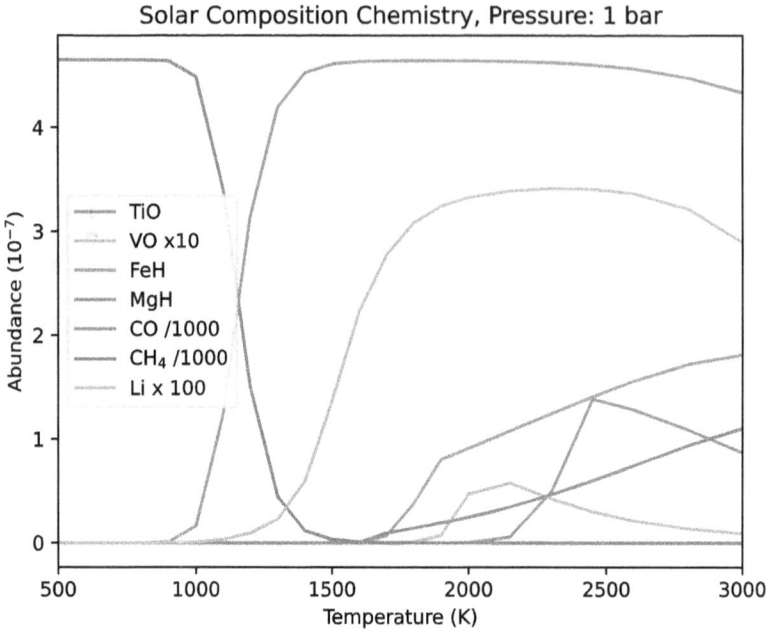

Figure 6.5. An illustration of chemical equilibrium results for selected molecules. Note the very different scales! The disappearance of molecules moving from hot (right) to cold (left) tracks the disappearance of features in the empirical L dwarf sequence. The T dwarfs are dominated by methane rather than CO as expected at lower temperatures.

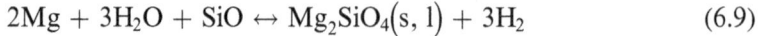

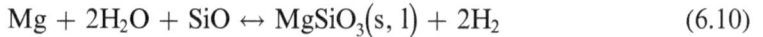

$$2Mg + 3H_2O + SiO \leftrightarrow Mg_2SiO_4(s, l) + 3H_2 \qquad (6.9)$$

$$Mg + 2H_2O + SiO \leftrightarrow MgSiO_3(s, l) + 2H_2 \qquad (6.10)$$

Let us consider what this means for magnesium, illustrated in Figure 6.6. Deep in the brown dwarf atmosphere, it is hot enough that these silicates do not condense. The elemental magnesium can be found there as atomic Mg or in molecules such as MgH, MgO, MgOH, etc., as found by solving the chemical reaction network. At some layer in the atmosphere, and higher, the atmosphere is cool enough (and low enough pressure) that the reactions above proceed and condensates are formed. Only $\sim 1\%$ of the magesium is left in the gas phase to form atoms or molecules. The gas phase has also been depleted in Si, so molecules involving Si are also rare, and in O, but because O is much more abundant than Mg, there is still plenty of O left over for CO, H$_2$O, and so forth. The key question is what happens to the condensates (Mg$_2$SiO$_4$, MgSiO$_3$). It is likely that the grains grow and settle out of the upper atmosphere into a cloud layer. The vertical thickness of the cloud and the size distribution of particles must be modeled. However, in most cases, it is clear that the cloud does not extend all the way to the top of the atmosphere, so that the upper atmosphere is depleted in Mg (and Si and O) even though the condensates are not

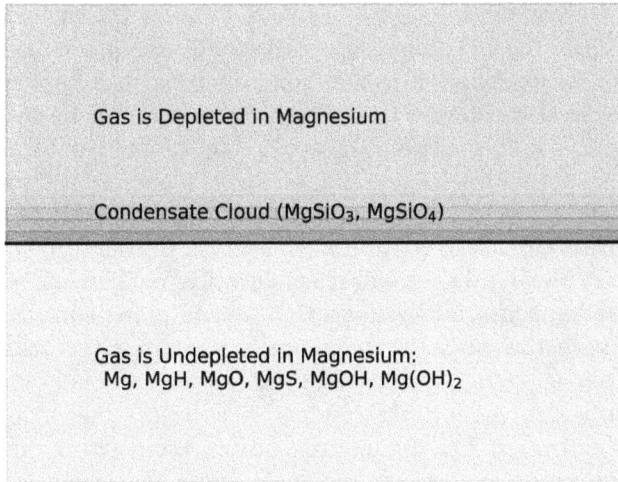

Figure 6.6. Illustration of atmospheric regions with condensate cloud formation and rainout chemistry involving magnesium. A silicate cloud layer is present where the condensates can condense extends above it with some finite thickness. Below the clouds, atomic and molecular magnesium is found but above the clouds magnesium is depleted, preventing formation of MgH and other molecules.

present. Calculations that include this effect for condensates are called "rainout chemistry" or "condensate cloud chemistry." Under rainout, the material will reach a different chemical equilibrium state. For example, in a proto-planetary disk, where the material remains mixed and does not rainout, the magnesium will also form $(MgFe)_2SiO_4$ (olivene) and $Ca_2MgSi_2O_7$ (akermanite) but in the rainout scenario, they do not form because Ca and Fe have already been removed by other condensates that form at higher temperature and greater depths. Lodders (2002) criticizes the term "rainout" as misleading and recommends instead "condensate cloud formation." We living on Earth think of rain as falling from the clouds down below, but here we mean that the condensates fall into the clouds from above. Also, many of the particles are solid (snowout chemistry?) rather than liquid. (Condensates that do fall from the clouds to below will rapidly return to the gas phase.) Brown dwarf models in chemical equilibrium all account for condensate cloud formation. In cloudless models, the condensates are assumed to sink or fall deep into the atmosphere and do not contribute opacity to the atmosphere. (This approximation works well for T dwarfs but poorly for L dwarfs.)

6.3.3 Mixing and Non-Equilibrium Chemistry

Observations of Solar System planets and brown dwarfs show that the assumption that each layer of the atmosphere is in chemical equilibrium is sometimes not correct. For example, CO was detected at $4.7\mu m$ in Gl 229B by Noll et al. (1997). This is surprising if the gases are in equilibrium: Noll et al. (1997) note that the spectra in that region are emitted from a photospheric layer with $T \approx 800$ K and

$P \approx 1$ bar, and as we see in Figure 6.4, we do not expect any significant CO at that temperature. Instead. the CO abundance corresponds to a much warmer ($T \geqslant 1250$ K) temperature. An explanation is that some material has been mixed up from below so that the layer we observe in not in chemical equilibrium. In particular, let's imagine that some CO molecules are transported upwards by diffusion from a layer with $T \approx 1300$ K, but with $\tau \gg 1$ so that we do not observe it directly, to a layer with $T \approx 800K$ and $\tau \approx 1$ that we can observe. A model for eddy diffusion was first developed to explain CO observed in Jupiter (Prinn & Barshay 1977). If the material is transported very slowly, then it will reach chemical equilibrium and the CO will not survive. If it happens quickly, then CO will be present in the layer that we observe. Quickly, in this case, the dynamical timescale is less than the chemical reaction timescale ($t_{\mathrm{dyn}} < t_{\mathrm{chem}}$). We need models to estimate both timescales. Equation (6.7) is a net reaction, the result of a series of other reactions that may follow different pathways. The limiting reaction in the network must be correctly identified to determine the chemical timescale. Note that chemical reactions will proceed more slowly at lower temperatures and pressure, so that the chemical timescales are longer in the upper atmosphere than in the deep atmosphere. The dynamical timescale is taken to be the result of diffusion through eddy motion— either convection, or some sort of atmospheric turbulence in radiative regions. One approach is to adopt a parameter K_{zz}, which describes the eddy diffusion. The dynamical timescale is then

$$t_{\mathrm{dyn}} = \frac{l^2}{K_{zz}} \tag{6.11}$$

If we take the length scale to be the pressure scale height (H), a highly uncertain approximation, then we are left with a diffusion coefficient K_{zz} (units of $\mathrm{cm^2\ s^{-1}}$) that could be predicted by a theory of convection or turbulent atmospheric motions but in effect is a free parameter and is a crude approximation of complex physical processes (Heng 2017). In any case, in the context of the model, the "quench point" where $t_{\mathrm{dyn}} = t_{\mathrm{chem}}$ is identified in the model, and layers above that take on the CO and CH_4 abundances of the quench point. An example of how the chemical profile is changed is shown in Figure 6.7 using the ATMO 2020 (Phillips et al. 2020) models for $T_{\mathrm{eff}} = 500$ K, $\log g = 4.0$. The corresponding synthetic spectra are shown in Figure 6.8. The chemical equilibrium calculation predicts that there is very little CO in the upper atmosphere in such brown dwarf. The models with weak ($K_{zz} = 10^5\ \mathrm{cm^2}$ $\mathrm{s^{-1}}$) or strong ($K_{zz} = 10^7\ \mathrm{cm^2\ s^{-1}}$). Compared to the JWST spectrum in Figure 2.3, we can clearly see that there is a distinct CO absorption at 4.7μm, consistent with the non-equilibrium chemistry predictions. Other molecules including NH_3 and PH_3 can also be treated in this way but have their own chemical timescales. Note that modeling this is again an iterative process, because the different chemical composition changes the opacities and the PT structure, which in turn changes the chemistry calculations.

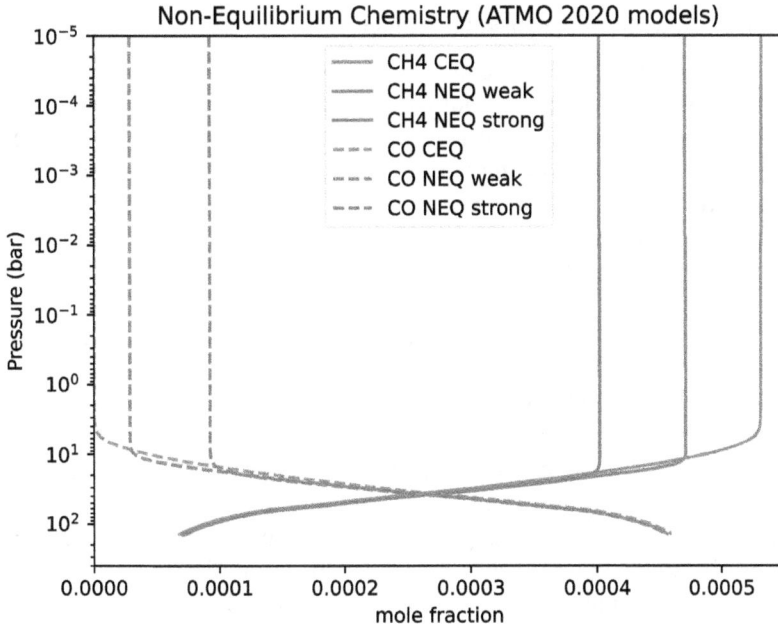

Figure 6.7. Carbon chemistry in the $T_{\mathrm{eff}} = 500$ K, $\log g = 4.0$ ATMO 2020 (Phillips et al. 2020) models for three cases: Chemical equilibrium (CEQ), "weak" non-equilibrium chemistry (NEQ-weak; $K_{zz} = 10^5$ cm^2 s^{-1}), and "strong" non-equilibrium chemistry (NEQ strong; $K_{zz} = 10^7$ cm^2 s^{-1}). CO is present in the upper atmosphere for the non-equilibrium cases.

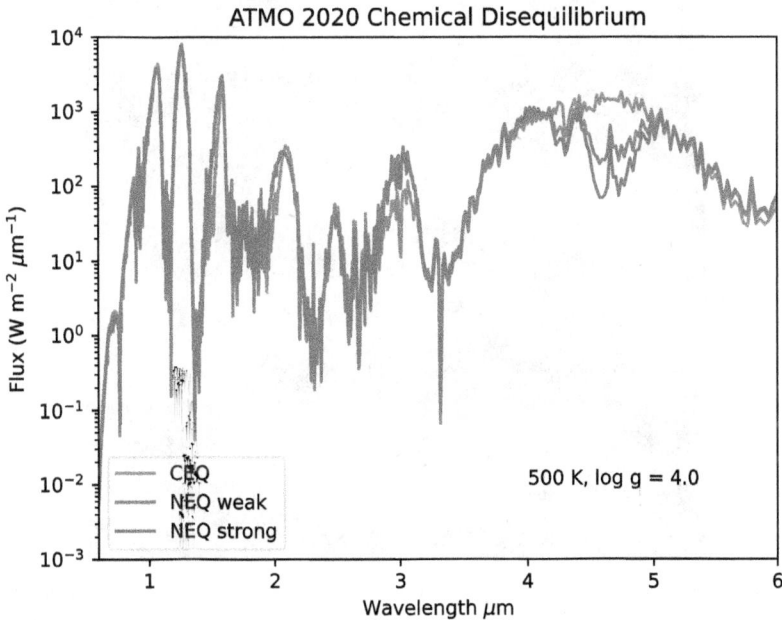

Figure 6.8. The spectra for the $T_{\mathrm{eff}} = 500$ K, $\log g = 4.0$ ATMO 2020 models (Phillips et al. 2020) corresponding to Figure 6.7. Note how the presence of CO suppresses the 4-5 μm ($W2$) peak.

6.3.4 Failure of the One-Dimensional Model

So far we have assumed that we can treat the brown dwarf's atmosphere as a one-dimensional problem (Figure 6.1), but let us to turn a Solar System planet to see how well this might work. Figure 6.9 shows an M-band (4.7 μm) image of Jupiter (Wong et al. 2020). The bright regions have gaps in the clouds where we can see deep into the atmosphere; hence, they appear as hotspots. These hotspots are dark in the visible spectrum, since the cloudy regions are better at reflecting sunlight. Observed as a distant point source, at 4.7μ m, Jupiter's spectrum would be dominated by these small hotspots.

Similar issues may occur with brown dwarfs. For example, it has been argued that the presence of FeH in mid-T dwarfs is not expected in rainout chemical equilibrium models, and that it is unlikely to be explained by vertical mixing and chemical disequilibrium models. Instead, the FeH may be due to regions where we see a deeper into the atmosphere, below the iron condensation line (Burgasser et al. 2002).

L and T dwarfs are typically photometrically and spectroscopically variable at $\sim$1% percent level indicating the presence of patchy clouds (Buenzli et al. 2014; Metchev et al. 2015), but at the L–T transition variability is frequent and much larger (>2%) as they pass through a transition from cloudy to clear with patchy cloud thickness variations (Apai et al. 2013; Radigan et al. 2014). Furthermore, as we noted in Chapter 3, there is evidence that viewing angles are correlated with

Figure 6.9. A 4.7μm (M-band) image of Jupiter taken with the Gemini North telescope. A one-dimensional atmospheric model is inadequate at this band and the observed light is dominated by hotspots where we can see into deep layers of the atmosphere. Creative Commons 4 license. International Gemini Observatory/ NOIRLab/NSF/AURA, M.H. Wong (UC Berkeley) et al. Acknowledgments: M. Zamani. CC BY 4.0.

observed cloud properties, indicating that the clouds in the equatorial region are different from the polar region.

6.4 Clouds

Tsuji et al. (1996) showed clouds are critically important for modeling late-M and L dwarfs. The relatively wavelength-independent opacity closes spectral windows, weakening the strength of absorption features. This opacity, through the greenhouse effect, warms the atmosphere, giving a warmer PT profile than cloud-free models. We've already seen that, in turn, it is important for the boundary conditions and influences the evolution of the star. To properly calculate the opacity, we need to know the size distribution and composition of the grains. We also need to know where the grains are located in the atmosphere. The upper atmosphere is cool, so that grains can condense there, but the models like DUSTY that have clouds all the way to the top of atmosphere for L dwarfs predict spectra that are too red with molecular features that are too weak. How can we characterize or predict the cloud distribution? Living on Earth, it should not be surprising that this is a very complex and challenging problem.

One approach is to model the microphysics of clouds in a self-consistent way: We might consider, for example, nucleation by TiO_2 (Sindel et al. 2022). As these first grains drift downwards in the atmosphere, they will grow. DRIFT-Phoenix is a model atmosphere family that includes the physics of grain growth and formation of cloud layers (Woitke & Helling 2004; Helling et al. 2008); a just-announced newer family is called MARCS-Static Weather-GGchem or MSG (Jørgensen et al. 2024). In Figure 6.10, we show the DRIFT-Phoenix synthetic spectra for late-M to mid-L dwarfs.[1] The agreement with the observed spectral types that we saw in Figure 3.5 is very good, and better than the cloudless models or the DUSTY models that had no settling, even though the models use older molecular opacity data. This particular model group, though, does not capture the L/T transition when clouds clear or sink below the photosphere, instead continuing to become redder. BT-Settl (Allard 2014) also incorporates a cloud microphysics and convection model. These models have no tunable free parameters, though no doubt choices were made in the cloud microphysics and convection models.

A different approach is the Ackerman & Marley (2001) cloud model. Informed by observations of Jupiter and more detailed cloud models, it attempts to capture cloud physics with a dimensionless free parameter (f_{sed} in the brown dwarf literature, f_{rain} in the original paper) that describes the efficiency of sedimentation. It is used in all the Sonora and earlier generation models. Here, the clouds are horizontally homogenous (appropriate for 1D atmosphere models) and their vertical extent is a balance of the downwards sedimentation rate, controlled by the free parameter, and the upwards turbulent mixing of the atmosphere. (Once again, we are modeling the rain or sedimentation into the cloud from above.) For smaller values of f_{sed}, the clouds extend higher vertically, but the particles are smaller. Most L dwarfs can be

---

[1] Phoenix is a model atmosphere code, also used for COND and DUSTY.

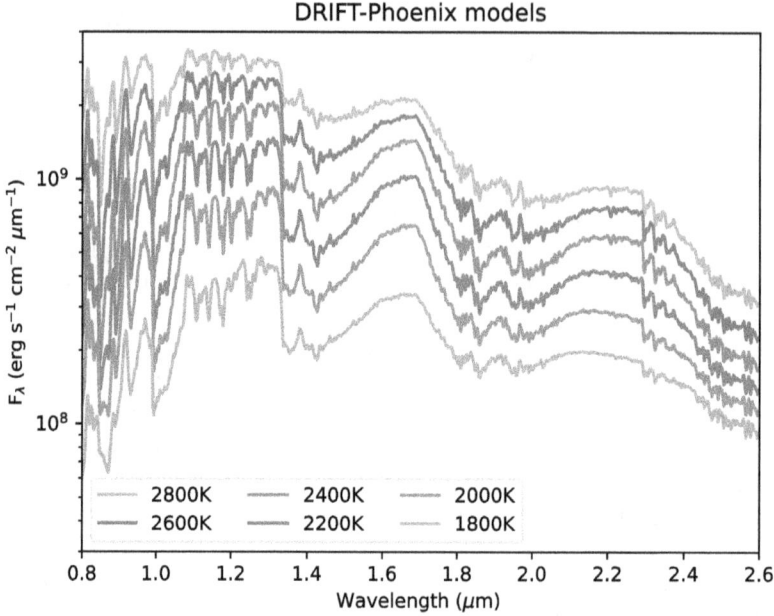

Figure 6.10. DRIFT-Phoenix models for 1800 K $\leqslant T_{\text{eff}} \leqslant$ 2800 K and $\log g = 5.0$. These models give a good description of late-M and early L dwarfs by modeling clouds.

fit with $f_{\text{sed}} = 2$, whereas models that include clouds for T dwarfs need a large $f_{\text{sed}} \geqslant 4$. Extremely red L dwarfs can produced using $f_{\text{sed}} = 0$ or 1.

6.5 Grid Models

There are now families of grid models that include all the parameters of interest for investigating both STAR-LIKE and SUPER-JUPITER research paradigms. Table 6.2 gives a selection of grid models for brown dwarfs—but this does not include all the important models from the last three decades. All model families include a range in T_{eff} and $\log g$. Available models include predictions for low metallicity, suitable for halo subdwarfs, and high metallicities for planets, ranges in C/O abundances, disequilibrium chemistry, and cloudy or cloudless models.

Although all models have some systematic discrepency with observations— always discussed in the model papers and subsequent uses of them—the current generations of models capture most of the characteristics of the spectral types we have observed. As a demonstration of the grid approach, in Figure 6.11, we fit the T8 dwarf GJ 570D to the cloudless, non-equilibrium chemistry ATMO 2020 NEQ-weak model family (Phillips et al. 2020). This brown dwarf has a precise Gaia parallax ($p = 169.88 \pm 0.07$ mas) so that the only free parameter in comparing the model $\mathcal{F}_\lambda$ to the observed F_λ is the radius $\mathcal{R}$: As in Equation (2.3), $f_\lambda = (\frac{\mathcal{R}}{d})^2 F_\lambda$. We use specutils to resample the models to the observed wavelengths (0.80μm

Table 6.2. Selected Grid Models

Name	Reference	Notes
COND	Allard et al. (2001)	Limiting case: clear
DUSTY	Allard et al. (2001)	Limiting case: thick clouds
DRIFT-Phoenix	Witte et al. (2011)	Self-consistent clouds
BT-Settl	Allard (2014)	Self-consistent clouds
M12	Morley et al. (2012)	L and T dwarf clouds
M14	Morley et al. (2014)	Water clouds
Exo-REM	Charnay et al. (2018)	Clouds, microphysics, planets
Sonora Bobcat	Marley et al. (2021)	Cloudless, chemical equilibrium
Sonora Cholla	Karalidi et al. (2021)	Cloudless, disequilibrium
Sonora Elfowl	Mukherjee et al. (2024)	Cloudless, disequilibrium, C/O
Sonora Diamondback	Morley et al. (2024)	Clouds
LB2023	Lacy & Burrows (2023)	Water clouds, disequilibrium
ATMO	Tremblin et al. (2015)	Fingering convection
ATMO 2020	Phillips et al. (2020)	Disequilibrium
ATMO 2020++	Meisner et al. (2023)	Adjusted adiabat
LOWZ	Meisner et al. (2021)	Cloudless low metallicity
SAND	Alvarado et al. (2024)	Clouds, low metallicity
MSG	Jørgensen et al. (2024)	Self-consistent clouds

$< \lambda < 2.4 \mu$ m) while conserving flux. We then consider $0.010 \, \mathcal{R}_\odot \leqslant \mathcal{R} < 0.150 \, \mathcal{R}_\odot$ in steps of $0.001 \, \mathcal{R}_\odot$, and for each $\mathcal{R}$ compute a fitting statistic, which is the sum of the squares of the difference between the observed spectrum and the model spectrum. We choose to weight all data points equally in this calculation. As we are simply demonstrating a grid search, we consider only four models, the three models with $T_{\mathrm{eff}} = 700$, 800, and 900 K and $\log g = 5.0$ plus one low-gravity model with $T_{\mathrm{eff}} = 800$ K and $\log g = 3.0$. The best fit model, in terms of our fitting statistic, is $T_{\mathrm{eff}} = 800$ K, $\log g = 5.0$, and $\mathcal{R} = 0.087 \, \mathcal{R}_\odot$, a very good match for our expectations for an old brown dwarf from evolutionary models even though our fitting procedure has no knowledge of this! The lower temperature model requires a larger radius and the high temperature a small radius, but both are distinctly worse fits, getting the relative strengths of the peaks wrong. We can also see that the $\log g = 3.0$ model is nearly as good, but it gets the shape of the H band wrong—and we would expect a larger radius for a very young, very-low-mass planetary mass object. Plainly one might make other choices in the fitting procedure: Should we weight by observational random noise or account for systematic calibration errors like an error in the slope? Should we include WISE or Spitzer mid-infrared photometry, or even a JWST mid-infrared spectrum if one becomes available? Should we discount regions of the spectrum where the model is known to have systematic issues? Should we use a sophisticated statistical test or perhaps machine learning? How should we interpolate the models to get a more

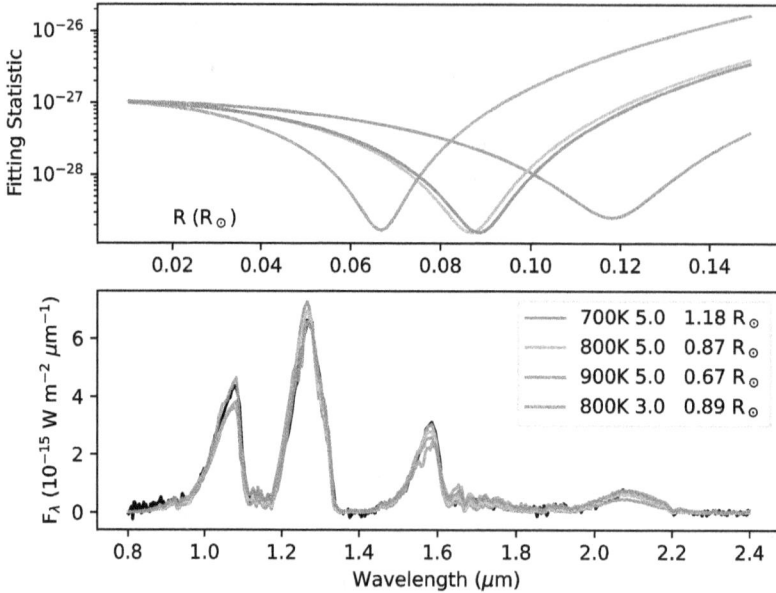

Figure 6.11. Example of fitting the T8 dwarf GJ 570D to the ATMO 2020 NEQ-Weak grid (Phillips et al. 2020). The observed SpeX spectrum is from Burgasser et al. (2004) and it has been flux calibrated to MKO $J = 14.820$ (Leggett et al. 2010). The system parallax of 169.88 so that the only free parameter in the fit is the radius $\mathcal{R}$. Top: The sum of the square of the differences between the observations and model for different assumed radii. Bottom: The best fit spectrum in each case. The overall best fit of the four shown is 800 K, $\log g = 5.0$ with $\mathcal{R} = 0.087\,\mathcal{R}_{\odot}$, which also matches our prior expectations from evolutionary models.

precise effective temperature? Nevertheless, the takeaway message is that current models have advanced to the point that we get apparently good parameters fits with a very simple fitting procedure. We will discuss some fundamental tests in the last chapter.

In summary, we can understand the main "extra" ingredients that are needed to capture the main observational trends. For M7 to mid-L dwarfs, including clouds is important. Different cloud properties can also explain the red and blue L dwarf peculiar classes. Metal-poor models capture the main trends seen in the halo subdwarfs. The L/T transition, where methane becomes prominent and $J - K$ colors shift from red to blue, requires a change in cloud properties. The one-dimensional approximation is likely inadequate here as large variability is observed. Tremblin et al. (2015) have advanced an alternative (ATMO) model for the L/T transition, where fingering convection and CO chemistry produce a much different PT profile without invoking clouds. T dwarfs need non-equilibrium chemistry, and while cloudless models fare well there may be some present. Our investigation of Y dwarfs is in the midst of being revolutionized by JWST, but data so far certainly require non-equilibrium chemistry while the role of water clouds is still being investigated. Changing the PT profile for Y dwarfs from the adiabatic convective predicted by standard energy transfer models improves the model fits (Leggett et al. 2021), and these "diabatic" profiles are available in ATMO 2020++ model grid. Retrieval methods are rapidly being developed and should eventually lead to reliable elemental abundances. This method is particularly important for the directly imaged exoplanets. In the next chapter, we turn to understanding the local solar neighborhood brown dwarf population under the STAR-LIKE research paradigm.

References

Ackerman, A. S., & Marley, M. S. 2001, ApJ, 556, 872

Allard, F. 2014, Exploring the Formation and Evolution of Planetary Systems IAU Symp., Vol. 299, IAU Symp., ed. M. Booth, B. C. Matthews, & J. R. Graham (Cambridge: Cambridge University Press) 271

Allard, F., Hauschildt, P. H., Alexander, D. R., Tamanai, A., & Schweitzer, A. 2001, ApJ, 556, 357

Allard, N. F., Kielkopf, J. F., Myneni, K., & Blakely, J. N. 2024, A&A, 683, A188

Alvarado, E., Gerasimov, R., Burgasser, A. J., et al. 2024, RNAAS, 8, 134

Apai, D., Radigan, J., Buenzli, E., et al. 2013, ApJ, 768, 121

Batalha, N. E., Marley, M. S., Lewis, N. K., & Fortney, J. J. 2019, ApJ, 878, 70

Buenzli, E., Apai, D., Radigan, J., Reid, I. N., & Flateau, D. 2014, ApJ, 782, 77

Burgasser, A. J., Geballe, T. R., Leggett, S. K., Kirkpatrick, J. D., & Golimowski, D. A. 2006, ApJ, 637, 1067

Burgasser, A. J., Marley, M. S., Ackerman, A. S., et al. 2002, ApJL, 571, L151

Burgasser, A. J., McElwain, M. W., Kirkpatrick, J. D., et al. 2004, AJ, 127, 2856

Burrows, A., & Sharp, C. M. 1999, ApJ, 512, 843

Charnay, B., Bézard, B., Baudino, J. L., et al. 2018, ApJ, 854, 172

Gharib-Nezhad, E., Marley, M. S., Batalha, N. E., et al. 2021, ApJ, 919, 21

Helling, C., Woitke, P., & Thi, W.-F. 2008, A&A, 485, 547

Heng, K. 2017, Exoplanetary Atmospheres: Theoretical Concepts and Foundations (Princeton, NJ: Princeton Univ. Press)

Hubeny, I., & Mihalas, D. 2014, Theory of Stellar Atmospheres (Princeton, NJ: Princeton Univ. Press)

Jørgensen, U. G., Amadio, F., Campos Estrada, B., et al. 2024, A&A, 690, A127

Karalidi, T., Marley, M., Fortney, J. J., et al. 2021, ApJ, 923, 269

Lacy, B., & Burrows, A. 2023, ApJ, 950, 8

Leggett, S. K., Burningham, B., Saumon, D., et al. 2010, ApJ, 710, 1627

Leggett, S. K., Tremblin, P., Phillips, M. W., et al. 2021, ApJ, 918, 11

Lodders, K. 2002, ApJ, 577, 974

Lodders, K. 2010, Principles and Perspectives in Cosmochemistry (Berlin: Springer-Verlag) 379

Lodders, K., & Fegley, B. 2002, Icar, 155, 393

Marley, M. S., Saumon, D., Visscher, C., et al. 2021, ApJ, 920, 85

Meisner, A. M., Leggett, S. K., Logsdon, S. E., et al. 2023, AJ, 166, 57

Meisner, A. M., Schneider, A. C., Burgasser, A. J., et al. 2021, ApJ, 915, 120

Metchev, S. A., Heinze, A., Apai, D., et al. 2015, ApJ, 799, 154

Mollière, P., Wardenier, J. P., van Boekel, R., et al. 2019, A&A, 627, A67

Mollière, P., Stolker, T., Lacour, S., et al. 2020, A&A, 640, A131

Morley, C. V., Fortney, J. J., Marley, M. S., et al. 2012, ApJ, 756, 172

Morley, C. V., Marley, M. S., Fortney, J. J., et al. 2014, ApJ, 787, 78

Morley, C. V., Mukherjee, S., Marley, M. S., et al. 2024, ApJ, 975, 59

Mukherjee, S., Batalha, N. E., Fortney, J. J., & Marley, M. S. 2023, ApJ, 942, 71

Mukherjee, S., Fortney, J. J., Morley, C. V., et al. 2024, ApJ, 963, 73

Noll, K. S., Geballe, T. R., & Marley, M. S. 1997, ApJL, 489, L87

Phillips, M. W., Tremblin, P., Baraffe, I., et al. 2020, A&A, 637, A38

Prinn, R. G., & Barshay, S. S. 1977, Sci., 198, 1031

Radigan, J., Lafrenière, D., Jayawardhana, R., & Artigau, E. 2014, ApJ, 793, 75

Sindel, J. P., Gobrecht, D., Helling, C., & Decin, L. 2022, A&A, 668, A35

Tennyson, J., & Yurchenko, S. N. 2017, IJQC, 117, 92

Tremblin, P., Amundsen, D. S., Mourier, P., et al. 2015, ApJL, 804, L17

Tsuji, T., Ohnaka, K., & Aoki, W. 1996, A&A, 305, L1

Visscher, C., Lodders, K., & Fegley, B. Jr. 2010, ApJ, 716, 1060

Witte, S., Helling, C., Barman, T., Heidrich, N., & Hauschildt, P. H. 2011, A&A, 529, A44

Woitke, P., & Helling, C. 2004, A&A, 414, 335

Wong, M. H., Simon, A. A., Tollefson, J. W., et al. 2020, ApJS, 247, 58

An Introduction to Brown Dwarfs
From very-low-mass stars to super-Jupiters
John Gizis

Chapter 7

The Solar Neighborhood

7.1 Brown Dwarfs Within 20 Parsecs and the IMF

The Galaxy has billions of brown dwarfs and planetary mass objects, but we are typically limited to observing only the relatively nearby ones. Having developed the tools to measure and predict $\mathcal{L}$ and T_{eff}, we can now turn to modeling the Solar Neighborhood brown dwarf population. We will define the Solar Neighborhood as a spherical volume centered on the Sun. The Gaia Catalog of Nearby Stars (Gaia Collaboration et al. 2021) lists nearly every hydrogen-burning star within 100 pc of the Sun, except for unresolved companions, but most ultracool dwarfs within that volume are too faint for Gaia. Instead, nearby brown dwarf samples have been built up by laborious followup of optical and infrared surveys since the discovery of confirmed brown dwarfs in 1995. Depending on the survey and spectral type considered the practical brown dwarf detection limits can approach 100 pc or be just a few parsecs. As of 2024, brown dwarf catalogs thought to be nearly complete are largely based on samples out to ~20 pc.

In a cylindrical Galactocentric frame, the Sun is at a radius $R = 8$ kpc and lies just $z = 20$ pc above the mid-plane. The brown dwarfs in the Solar Neighborhood represent a random sampling of those formed over the full history of the Galaxy over a wide range of positions within the disk. To illustrate this, we use galpy (Bovy 2015) to calculate the last 100 Myr of motion of the ten apparently single nearby L dwarfs within 10 pc whose parallaxes, proper motions, and radial velocities have been measured. Figure 7.1 shows that these L dwarfs were not close to the Sun in the past. All, however, are members of the Galactic Disk population, remaining within a few hundred parsecs of the Galactic Plane even as they move more than a kiloparsec in radius. The local sample can be taken to be a fair sampling of the Galactic disk population, except that we know that the nearest star-forming regions are >100 pc, so there are very few stars with ages <10 Myr, and the young ($\lesssim 100$ Myr) may not be well mixed.

doi:10.1088/2514-3433/ad757ech7   7-1

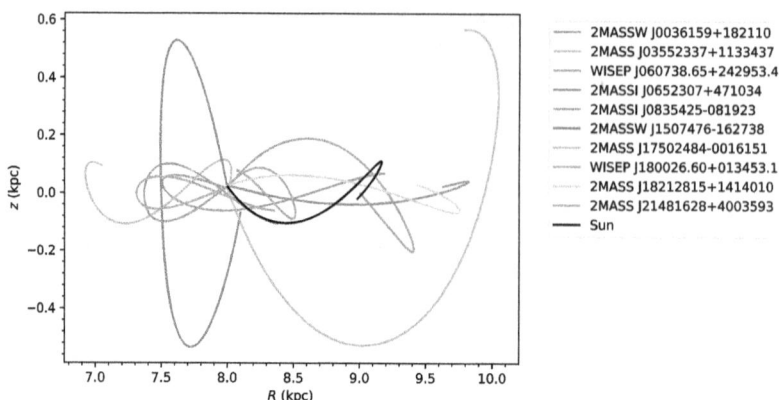

Figure 7.1. The position of 10 nearby L dwarfs over the last 100 Myr.

Galactic Orbits The galpy Python package (Bovy 2015) allows us to numerically integrate orbits in our Galaxy. It is available at http://github.com/jobovy/galpy

Armed with brown dwarf evolutionary models, we therefore consider how we can model the local Solar Neighborhood population of ultracool dwarfs. To illustrate the problem, we begin with what we will call the Simple model of this solar population. We'll consider only the Galactic Disk (or Population I) population. We will assume the Disk formed 10 Gyr ago and assume the star formation rate was uniform, so that all ages (0.01–10 Gyr) are equally likely. Our Simple model excludes the youngest brown dwarfs (<10 Myr) and the Thick Disk and Halo populations of very old brown dwarfs, assuming for now that such objects could be identified and counted separately, but in any case, these are a small percentage of objects. Although many M, L, and T subdwarfs have been identified, Kirkpatrick et al. (2024) point out that the 20 parsec sample has only two ultracool subdwarfs, the esdL1 dwarf SSSPM J1144-2019 (Scholz et al. 2004; Zhang et al. 2018) and the Y subdwarf The Accident (Kirkpatrick et al. 2021), that can be assigned to the true halo. This is consistent with the expected 0.2%.

We draw masses from a power-law mass distribution:

$$\frac{dN}{d\mathcal{M}} \propto \mathcal{M}^{-\alpha} \tag{7.1}$$

Notice that if $\alpha = 0$, the number of objects in equal mass bins would be equal, so that there would be the same number of brown dwarfs in the 0.02 $\mathcal{M}_\odot$-0.03 $\mathcal{M}_\odot$ as stars in the 0.090 $\mathcal{M}_\odot$ − 0.100 $\mathcal{M}_\odot$; and this would be the same as the number of objects between 0.003 $\mathcal{M}_\odot$ and 0.013 $\mathcal{M}_\odot$! Overall, there would be six times more objects in the 0.013 − 0.073 brown dwarf mass range than the 0.002–0.012 $\mathcal{M}_\odot$ planetary mass range. On the other hand, for $\alpha = 1$, the number of objects is constant for logarithmic mass bins, so that the number of planetary mass objects in the range 0.003–0.012 $\mathcal{M}_\odot$ –a factor of four in mass—would equal the number of brown dwarfs from 0.012 to 0.048 $\mathcal{M}_\odot$. Recent analysis of the Solar Neighborhood sample suggests $\alpha \approx 0.6$ and we will adopt this for the illustrative models in this chapter.

In the Simple model, we adopt $\alpha = 0.6$ from a minimum mass of 0.005 $\mathcal{M}_\odot$ to a maximum mass of 0.080 $\mathcal{M}_\odot$. Once we draw a random age and mass for each simulated object, we use the Hybrid cloud Saumon & Marley (2008) solar metallicity models to predict its luminosity and effective temperature. The Simple model is attractive but has clear drawbacks compared to our knowledge of stars and the Galaxy. First, we know that older Galactic disk stellar populations are more extended vertically: Stars are born in kinematically cold, molecular clouds near the disk mid-plane, and then are gradually scattered to hotter kinematics and higher heights above and below the plane by encounters with massive perturbers. Put another way, we observe the local space density (ρ in units of numbers or solar masses per cubic parsec) but a full accounting would integrate over all heights to get the local surface density (Σ in units of numbers or solar masses per square parsec) for each population i:

$$\Sigma_i = \int_{-\infty}^{\infty} \rho_i(z) \, dz \qquad (7.2)$$

The effects of molecular clouds, spiral arms, and other massive objects dominate over star-star scattering, so that brown dwarfs should show the same age dependence as the stars.[1] The successful TRILEGAL star count model Girardi et al. (2005) models the vertical distributions as a $\sinh^2$ function:

$$\rho(z) \propto 0.25 \sinh^2 (0.5z/h) \qquad (7.3)$$

The scale height is well described by

$$h = z_0(1 + t/t_0)^a \qquad (7.4)$$

The TRILEGAL model uses parameters $z_0 = 94.6902$ pc, $t_0 = 5.550\,79 \times 10^9$ yr, and $a = 1.6666$ providing the best fit to star counts. The effect on the local densities is illustrated in Figure 7.2: Brown dwarfs that are one billion years old are more than twice as likely to be near the Sun as those that are five billion years old![2]

The age effect can also be seen in the local space velocities (Figure 7.3), with U in the direction of the Galactic center, V in the direction of Galactic Rotation, and W perpendicular to the disk. We adopt the model Constant which has constant star formation but takes account of the vertical scale height as a function of age. We also

---

[1] Whether the scale heights and kinematics of the oldest disk objects are best explained by this scattering mechanism, or other scenarios such as mergers with small galaxies or a more extended gas distribution is an important problem, but does not matter for our purposes.

[2] The Besançon Galaxy model (Czekaj et al. 2014) does not use analytic models for the vertical distribution but instead self consistently solves for the vertical mass distribution. We use the TRILEGAL analytic model for convenience, but see Dupuy & Liu (2017) for Besançon applied to brown dwarfs.

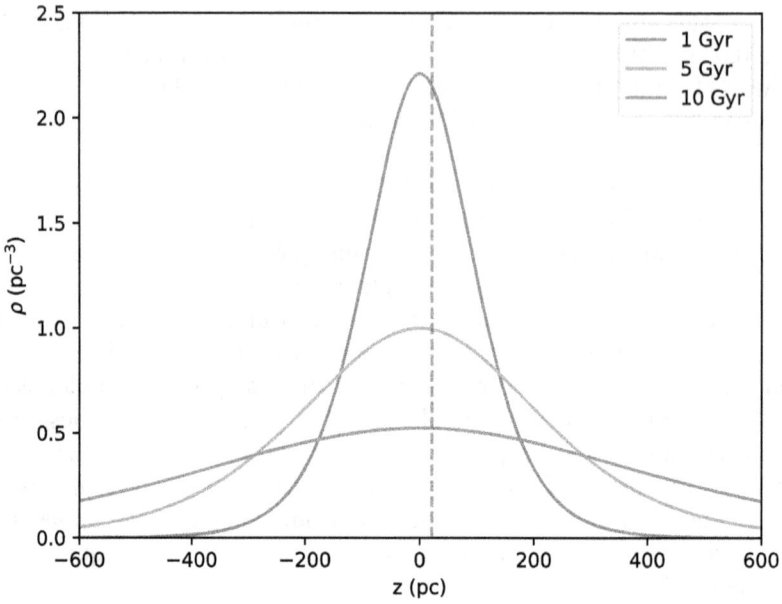

Figure 7.2. The stellar density distribution perpendicular to the Galactic plane as a function of age (Girardi et al. 2005). The position of the Sun ($z \approx 21$ pc) is marked. Young brown dwarfs will be overrepresented in the Solar Neighborhood compared to older ones.

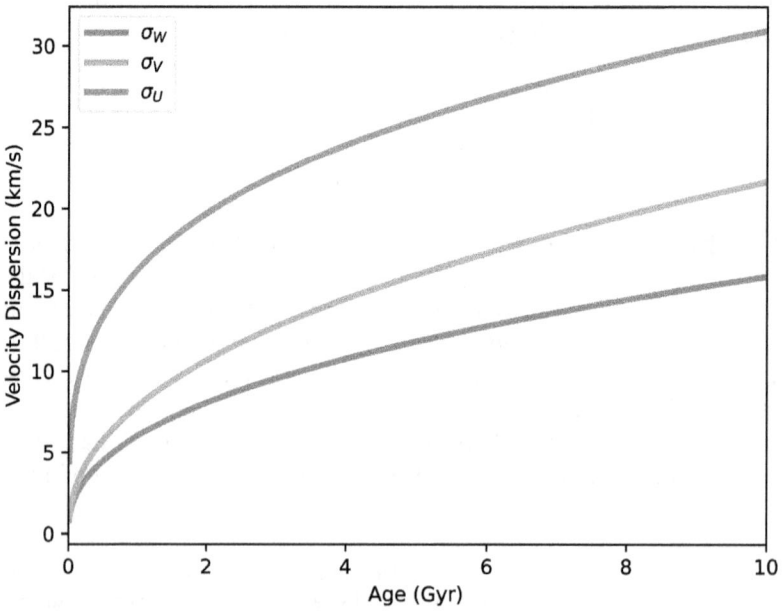

Figure 7.3. The local stellar velocity dispersions as a function of age (Crandall et al. 2022). Young brown dwarfs will be kinematically cooler than an old population.

Star Formation History

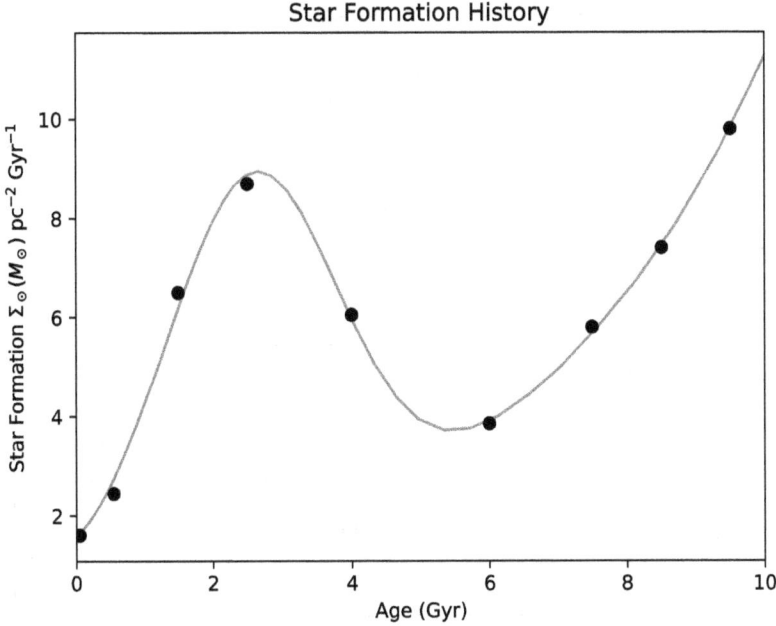

Figure 7.4. The Galactic Disk Star Formation history according to Mor et al. (2019), which we use for our Bursty Solar Neighborhood model. Our fitted values are $\Sigma = \Sigma_0 e^{\gamma t} + \Sigma_1 e^{(t-\mu)^2/2\sigma^2}$. $\Sigma_0 = 0.721$, $\gamma = 0.275$, $\Sigma_1 = 7.467$, $\mu = 2.575$, $\sigma = 1.226$.

assign a random W velocity in our simulations from the disk-age relation shown in Figure 7.3. The second issue is that it is unlikely that the star formation history of the Galactic disk has been constant. Models and observations of disk galaxies tend to find that the star formation was exponentially higher in the past ($e^{\gamma t}$), some have found that the Galactic disk star formation rate (i.e., $d\Sigma/dt$) is consistent with $\gamma = 0.12$ Gyr^{-1} (Czekaj et al. 2014; Mor et al. 2018). We present this model (exponential decline in $d\Sigma/dt$) as model decaying. However, evidence in favor of a burst of star formation about 2–3Gyr ago has been accumulating with new Gaia data (Isern 2019; Mor et al. 2019; Mazzi et al. 2024; Gallart et al. 2024). We adopt the Mor et al. (2019) star formation history, which can be described as an exponential plus a Gaussian burst (Figure 7.4). Note that this model has a low rate of recent (<0.5 Gyr) star formation. These models are meant to illustrate a wide range of plausible possibilities for the Solar Neighborhood (Figures 7.5 and 7.6) and the exciting possibilities as the star formation history becomes better constrained.

We can compare our simulations to actual space densities of ultracool dwarfs. First, let us consider the procedure for measuring this. If we had a complete, full-sky survey of all objects with trigonometric parallaxes complete out to a distance of 20 pc, then clearly the number of objects with, for example, bin i with $1200 < T_{\text{eff}} < 1350$ K would the number of such objects divided by the volume $(4\pi(20\text{pc})^3/3)$, and of course we would simply reduce the $4\pi = A$ factor if we covered less area than the full sky. Imagine, however, that we have many objects

Solar Neighborhood Age Distribution

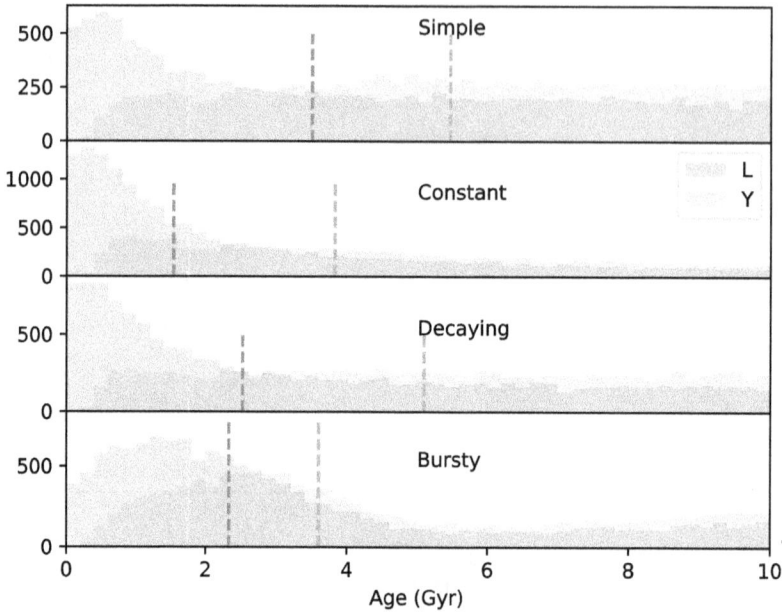

Figure 7.5. Age distributions for $1400 < T_{eff} < 2200$ K (labeled as L dwarfs) and $350 < T_{eff} < 450$ K (Y dwarfs). The corresponding median values are marked with a dashed line. Note that even in the `Simple` model where the mean and median age of all stars and brown dwarfs is 5 Gyr, the L dwarf population is significantly younger. These models all have IMF slope $\alpha = 0.6$.

that lie in that temperature bin, but each could be seen only to some distance less than 20 pc depending upon its absolute magnitude and the magnitude limit of the survey, and we covered A steradians on the sky. In that case, for each object detected, we calculate the maximum distance it could have been detected at (d_{max}), and the corresponding volume it could have been seen is $V_{max} = A d_{max}^3 / 3$. In this case, the maximum likelihood estimator of the space density for the ith population is simply

$$\rho_i = \sum \frac{1}{V_{max}} \tag{7.5}$$

We could also apply incompleteness corrections, if, for example, if we believe 10% of objects are missed for some observational reason, or use photometric distance estimates if necessary. The luminosity function for the solar neighborhood is the count of dwarfs per cubic parsec per absolute magnitude, either bolometric or through some filter. The mass function for the solar neighborhood is the count of dwarfs per cubic parsec per solar mass, or equivalent. Two groups—Best et al. (2021) and Kirkpatrick et al. (2024)—each have measured a variant on the luminosity function: the number of dwarfs per cubic parsec per 150K effective temperature, or equivalent, based on dwarfs within 15–25 parsecs. The two estimates

Solar Neighborhood Mass Distribution

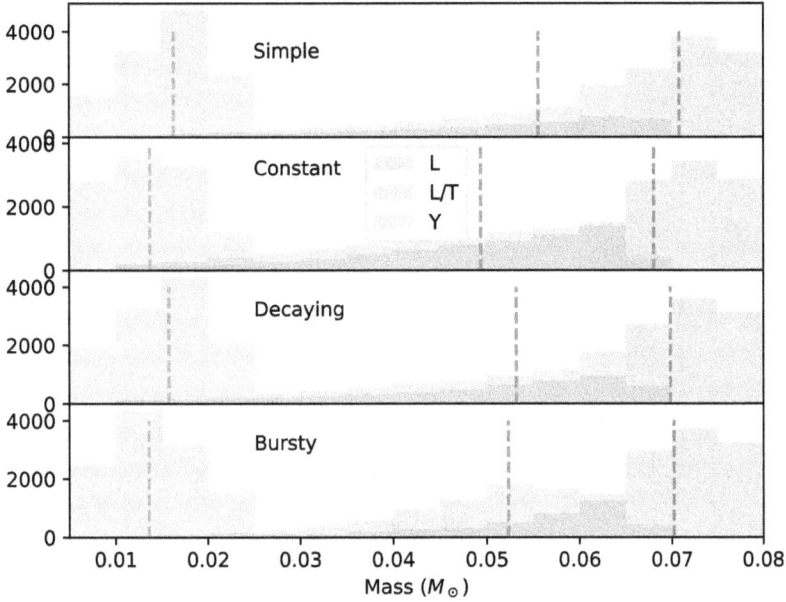

Figure 7.6. Mass distributions for $1400 < T_{\mathrm{eff}} < 2200$ K (labeled as L dwarfs), $1250 < T_{\mathrm{eff}} < 1350$ K (L/T transition), $350 < T_{\mathrm{eff}} < 450$ K (Y dwarfs). The corresponding median values are marked with a dashed line. These models all have IMF slope $\alpha = 0.6$.

are similar—the samples overlap and much of the underlying data is the same—but involve different analysis choices. We plot the Kirkpatrick et al. (2024) space densities in Figure 7.7 for three cases: The Simple model with the Hybrid (Saumon & Marley 2008) models, the Simple model with cloudless Sonora Bobcat (Marley et al. 2021) models, and a Bursty model with IMF $\alpha = 0.8$ and Hybrid evolution. The observed peak for $1200 < T_{\mathrm{eff}} < 1350$ K is only matched by the Hybrid models and is the best evidence that a fairly abrupt transition is needed to explain the otherwise surprisingly large number of L/T transition objects. The power-law mass function slope $\alpha = 0.6$ is in good agreement with the data, not surprisingly because this is essentially the same model that Kirkpatrick et al. (2024) used to fit $\alpha = 0.6$. The Bursty agrees better with $\alpha = 0.8$, illustrating that the uncertainty due to the star formation history is α is ± 0.2. For comparison, Best et al. (2024) used a model with a simple geometry but allowed the star formation rate (γ in our language) to vary to account for both star formation and scale heights, finding $\alpha = 0.58^{+0.16}_{-0.20}$ and $\gamma = -0.44 \pm 0.4$. There is, in short, very little doubt that the Simple model does not agree with our knowledge of the Galactic stellar population or the observed properties of the brown dwarfs. That said, the substellar IMF need not be a simple power law and the star formation history of the Solar Neighborhood remains uncertain.

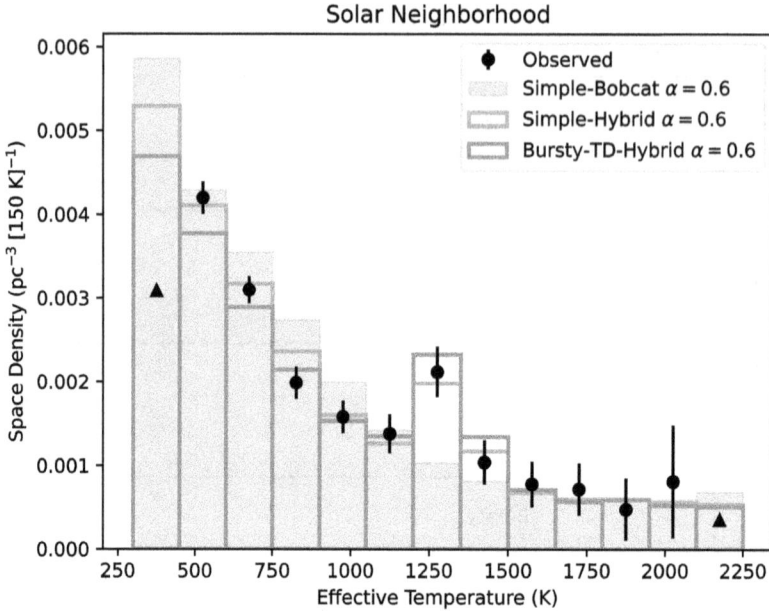

Figure 7.7. The observed space densities in the Solar Neighborhood as a function of effective temperature compared to our `Simple` and `Bursty` models. The lack of Y dwarfs with ages less than 300 Myr is due to the minimum mass cutoff of $0.005\,\mathcal{M}_\odot$. Bobcat and other cloudless models do not predict the peak around the L/T transition at $1200 - 1350$ K. We include a thick disk with a local normalization of 2% with Sonora Bobcat $[M/H] = -0.5$ models with ages 9–10 Gyr.

7.2 Very Young Stars and Nearby Moving Groups

The Sun sits in the Local Bubble of the interstellar medium—very convenient for being able to neglect extinction in our brown dwarf samples, but as a result, there are very few young (<20 Myr) stars within 100 pc compared to their space densities from 100 to 500 pc (Zari et al. 2018). Perhaps the most famous exception is the T Tauri star TW Hya (Rucinski & Krautter 1983), which is far from any dark cloud. More than two dozen stars, brown dwarfs and planetary mass objects are now known to be co-moving with TW Hya (Gagné et al. 2017), known as the TW Hya Association (TWA). Most importantly, members of a moving group or association can be compared to each other and their age jointly estimated. Today, numerous moving groups are well established: The Banyan-Σ tool assesses the probability of membership in 27 different associations within 150 pc with ages from 1 to 800 Myr, ranging from rich open clusters like the Hyades and the Pleiades to groups with seven members as 2018, and a huge number of associations, tidal tails, and other kinematic substructures continue to be identified using Gaia.[3] Most remarkably, the new Oceanus Moving Group (Gagné et al. 2023) may include Luhman 16. Some

---

[3] On a larger scale, over 13,000 Galactic open clusters, more substantial but usually much more distant, are now listed in the Unified Cluster Catalog (Perren et al. 2023).

Table 7.1. Selected Nearby Moving Groups

Name	Age (Myr)	Age Reference
TW Hya (TWA)	10 ± 3	Bell et al. (2015)
β Pic (BPMG)	24 ± 3	Bell et al. (2015)
Columba	42^{+11}_{-7}	Bell et al. (2015)
Carina	45^{+11}_{-7}	Bell et al. (2015)
Tuc-Hor	45 ± 4	Bell et al. (2015)
AB Dor (ABDMG)	149^{+51}_{-19}	Bell et al. (2015)
Oceanus	~ 500	Gagné et al. (2023)
Hyades	650 ± 70	Lodieu (2020)

selected moving groups particularly important for brown dwarf research are listed in Table 7.1, but this is a highly incomplete list. Sky surveys have reached near to and below the deuterium-burning limit for many of these moving groups, and they have also been heavily targeted with high-resolution direct imaging searches for young gas-giant planets. The resulting discoveries have enabled the low-gravity spectral type classification systems (Chapter 3) and to explore STAR-LIKE and SUPER-JUPITER scenarios.

As an example: The Gemini Planet Imager (GPI) Exoplanet Survey (Nielsen et al. 2019), a survey that targeted 300 nearby young stars, mostly in moving groups, found six companions below the deuterium-burning limit and three brown dwarfs. The companion rate can be characterized as a power-law function of companion mass ($\mathcal{M}$) but also needs to account for the semimajor axis (a) and primary star mass (M_*)

$$\frac{dN}{dm\,da} = f C_1 \mathcal{M}^{-\alpha} a^\beta \left(\frac{\mathcal{M}_*}{1.75\,\mathcal{M}_\odot} \right)^{\gamma'} \qquad (7.6)$$

However, unlike interpreting the solar neighborhood space densities, fitting these parameters requires a model of the sensitivity to companions as a function of separation on a target-by-target basis so we will not attempt to develop our own models. (Note that compared to the original paper, we have changed the sign of α to be consistent with our IMF Equation (7.1), and we use γ' for the stellar mass dependence to distinguish from our earlier use of γ in this chapter.) For brown dwarfs ($13\,\mathcal{M}_{\mathrm{Jup}} < \mathcal{M} < 80\,\mathcal{M}_{\mathrm{Jup}}$), the authors find $\alpha \approx 0.5$, $\beta \approx -0.7$, $\gamma \approx -0.9$. Only $0.8^{+0.8}_{-0.5}\%$ of stars have a brown dwarf companion. In contrast, 'planets' ($5\,\mathcal{M}_{\mathrm{Jup}} < \mathcal{M} < 13\,\mathcal{M}_{\mathrm{Jup}}$) are more frequent ($9^{+5}_{-4}\%$), and more strongly favor lower masses and closer separations $\alpha \approx 2.3$, $\beta \approx -1.7$, $\gamma \approx 2$. Similarly, although adopting different functional forms for the companion separation model, Reggiani et al. (2016) find that the outcome of a large VLT survey also favors two populations: A brown dwarf companion fraction that is only 1% between 28 and

1590 AU for solar-type stars and that extends down to the (assumed) fragmentation limit of 5 $\mathcal{M}_{\mathrm{Jup}}$, and a planet population that strongly increases to lower masses ($\alpha \approx 1.3$).

References

Bell, C. P. M., Mamajek, E. E., & Naylor, T. 2015, MNRAS, 454, 593

Best, W. M. J., Liu, M. C., Magnier, E. A., & Dupuy, T. J. 2021, AJ, 161, 42

Best, W. M. J., Sanghi, A., Liu, M. C., Magnier, E. A., & Dupuy, T. J. 2024, ApJ, 967, 115

Bovy, J. 2015, ApJS, 216, 29

Crandall, S., Smith, G. H., Birmingham, S., et al. 2022, Ap&SS, 367, 50

Czekaj, M. A., Robin, A. C., Figueras, F., Luri, X., & Haywood, M. 2014, A&A, 564, A102

Dupuy, T. J., & Liu, M. C. 2017, ApJS, 231, 15

Gagné, J., Moranta, L., Faherty, J. K., et al. 2023, ApJ, 945, 119

Gagné, J., Faherty, J. K., Mamajek, E. E., et al. 2017, ApJS, 228, 18

Gaia CollaborationSmart, R. L., & Sarro, L. M. 2021, A&A, 649, A6

Gallart, C., Surot, F., Cassisi, S., et al. 2024, A&A, 687, A168

Girardi, L., Groenewegen, M. A. T., Hatziminaoglou, E., & da Costa, L. 2005, A&A, 436, 895

Isern, J. 2019, ApJL, 878, L11

Kirkpatrick, J. D., Marocco, F., & Gelino, C. R.The Backyard Worlds: Planet 9 Collaboration 2024, ApJS, 271, 55

Kirkpatrick, J. D., Marocco, F., Caselden, D., et al. 2021, ApJL, 915, L6

Lodieu, N. 2020, MmSAI, 91, 84

Marley, M. S., Saumon, D., Visscher, C., et al. 2021, ApJ, 920, 85

Mazzi, A., Girardi, L., Trabucchi, M., et al. 2024, MNRAS, 527, 583

Mor, R., Robin, A. C., Figueras, F., & Antoja, T. 2018, A&A, 620, A79

Mor, R., Robin, A. C., Figueras, F., Roca-Fàbrega, S., & Luri, X. 2019, A&A, 624, L1

Nielsen, E. L., De Rosa, R. J., Macintosh, B., et al. 2019, AJ, 158, 13

Perren, G. I., Pera, M. S., Navone, H. D., & Vázquez, R. A. 2023, MNRAS, 526, 4107

Reggiani, M., Meyer, M. R., Chauvin, G., et al. 2016, A&A, 586, A147

Rucinski, S. M., & Krautter, J. 1983, A&A, 121, 217

Saumon, D., & Marley, M. S. 2008, ApJ, 689, 1327

Scholz, R.-D., Lodieu, N., & McCaughrean, M. J. 2004, A&A, 428, L25

Zari, E., Hashemi, H., Brown, A. G. A., Jardine, K., & de Zeeuw, P. T. 2018, A&A, 620, A172

Zhang, Z. H., Galvez-Ortiz, M. C., Pinfield, D. J., et al. 2018, MNRAS, 480, 5447

An Introduction to Brown Dwarfs
From very-low-mass stars to super-Jupiters
John Gizis

Chapter 8

Conclusions

8.1 Binaries as Tests of Models

We mainly discussed brown dwarfs as single objects. Brown dwarfs in binaries—and higher order systems—offer the potential to measure much more information and thereby constrain or test models. Kepler's Third Law gives us

$$(M_1 + M_2) = \frac{a^3}{P^2} \tag{8.1}$$

For a binary system, there are seven orbital parameters that need to be fitted. Four describe the physical properties of the orbit: The orbital period (P), the semimajor axis of the orbit (a, in arcseconds or au if the distance is known), the eccentricity (e), and the time of periastron passage (T). Three describe the geometry of our line of sight: The inclination angle (i), the position angle of a nodal point where the plane of the sky intersects the plane of the orbit (Ω), and the angle between the line of nodes and the semimajor axis (ω). Observations over time measure the angle (θ), separation ρ, and time of observation (t).

Let us start with one example: Brown dwarfs that are distant companions to higher mass stars. Here, we expect little measurable orbit motion—for $a \approx 100$ au and $M_1 + M_2 \approx 1\,\mathcal{M}_\odot$, $P \approx 1000$ yr. The brown dwarf and primary star will share the same proper motion, allowing them to be identified as a bound pair. The properties of the primary constrain the brown dwarf companion. Its Gaia parallax immediately gives us the distance of the brown dwarf without a prolonged parallax campaign. If the primary is an FGK star, then the standard analysis of a high-resolution optical spectrum will yield the element abundances relative to the Sun; even for M dwarfs, measurement of elemental abundances is becoming possible. We may have age constraints for the primary, such as from isochrone fitting, magnetic activity-age relationships, or rotation period-age (gyrochronology).

doi:10.1088/2514-3433/ad757ech8

GJ 570D (Burgasser et al. 2000) is a T8 brown dwarf that is in a wide orbit around a K4 primary (GJ 570A or HD 131 977) and a pair of M dwarfs (GJ 570BC). The Hypatia Catalog reports that the primary has [Fe/H] = 0.04, but Montes et al. (2018) measure [Fe/H] = −0.12 and Luck (2018) measure [Fe/H] = +0.12, illustrating the uncertainty even for relatively well-understood stars. Luck (2018) have an extensive analysis of GJ 570A that the abundances (C/O =0.54, solar), and includes isochrone fitting consistent with fairly old ages 5–8 Gyr. We have already seen in Chapter 6 that the SpeX spectrum of GJ 570D is well fit by T_{eff} = 800 K ATMO 2020 model. This same spectrum has been the subject of detailed grid fitting and retrieval fitting because of its interest as a benchmark system and its relative proximity. Phillips et al. (2024) fit both the SpeX spectrum and a higher resolution near-infrared spectrum with an updated ATMO 2020 grid that includes [M/H] and C/O as additional parameters. Figure 8.1 shows the results for analyses that fit T_{eff}, $\log g$, [M/H], and C/O. While the updated ATMO 2020 grid fitting yields solar C/O, near solar metallicity, and $\log g$ = 4.4 ± 0.25, very different results are found by retrieval analysis. Line et al. (2017), Kitzmann et al. (2020), Zalesky et al. (2022) and Whiteford et al. (2023) all find high C/O ratios ∼0.8 and most find cooler temperatures T_{eff} ≈ 720. The tension between these results is significant: in a STAR-LIKE scenario we expect the C/O to match the primary, as it does in the updated ATMO 2020 fitting. Perhaps the retrieval results are not yet reliable, or else

GJ 570D Fits

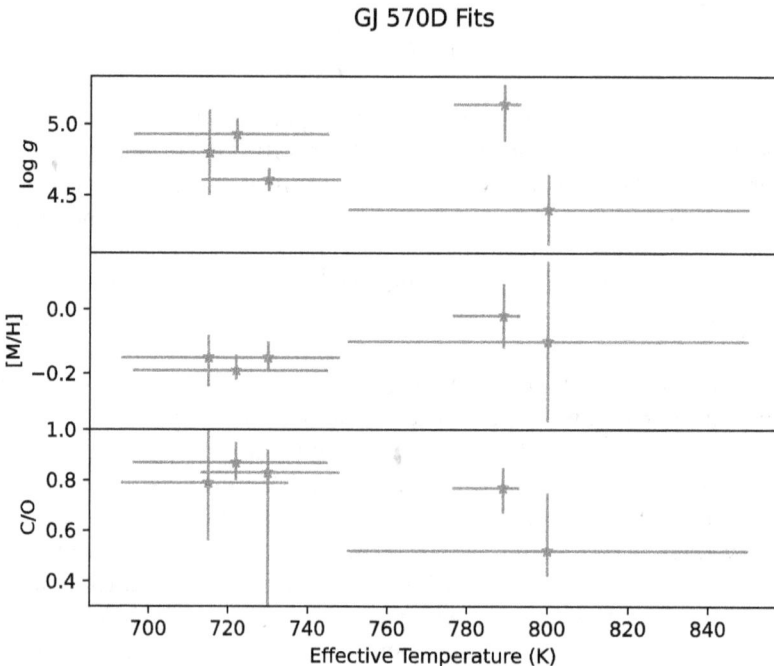

Figure 8.1. Results of independent analysis of the SpeX prism spectrum of GJ 570D. The updated ATMO 2020 results by Phillips et al. (2024) are the ones with T_{eff} = 800 ± 50 K. The others are the results of retrieval analysis. See the text for discussion.

there is something very surprising about GJ 570D's atmosphere or formation. JWST spectra covering the 1–12 μm range would be important data to check both the grid and retrieval approaches.

At the opposite extreme of separation, very close brown dwarf companions have a chance to be transiting their stellar primary. These are naturally identified in transit surveys that look for hot and warm Jupiters with $\mathcal{R} \approx 1$ $\mathcal{R}_{\mathrm{Jup}}$, and their mass can be measured with radial velocity followup. The depth of the transit is $(\mathcal{R}_{\mathrm{BD}}/\mathcal{R}_*)^2$, and Gaia parallaxes and often spectroscopy, the primary stars's $\mathcal{M}$, $\mathcal{R}$, and metallicity are well constrained. In Figure 8.2, we see the measured values from the Carmichael (2023) catalog of transiting brown dwarfs. Although many of the brown dwarfs are expected to be inflated by their close, hot primary stars, the lower envelope of the observed mass–radius distribution is consistent with standard interior evolutionary models, with perhaps one outlier. This is a success for the theory of very-low-mass stars and brown dwarfs.

A decade-long monitoring campaign by Dupuy & Liu (2017) measured the orbits and photocenters of nearby ultracool dwarf binaries, yielding both the total system and individual masses. The measured masses as a function of observed spectral type are shown in Figure 8.3. Also shown are two benchmarks companions to nearby K

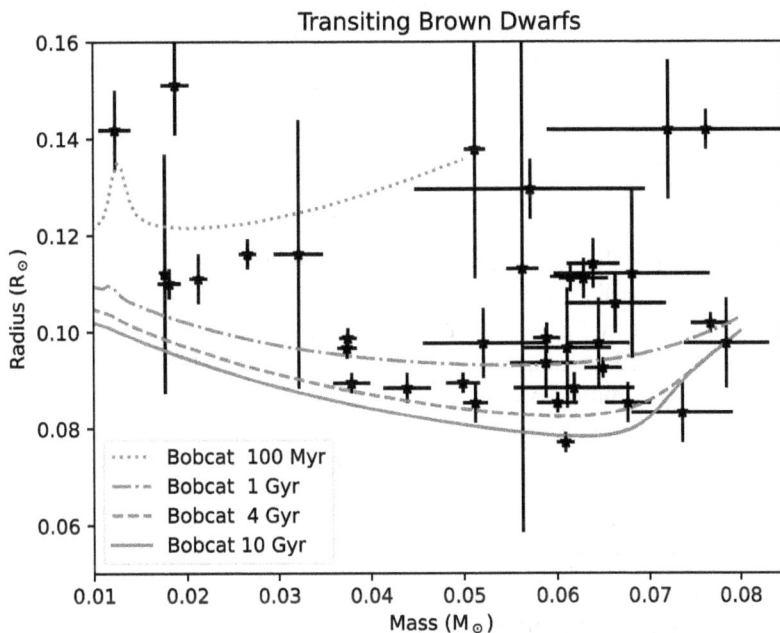

Figure 8.2. Measured mass and radius for transiting brown dwarfs analyzed and compiled by Carmichael (2023) compared to the Sonora Bobcat solar-metallicity models (Marley et al. 2021). Many of these are close to the primary star and may be inflated due to irradiation, but for the most part the brown dwarfs are consistent with the model predictions. The smallest radius object is TOI-569b with $\mathcal{M} = 63.8 \pm 1.0$ $\mathcal{M}_{\mathrm{Jup}}$ and $\mathcal{R} = 0.75 \pm 0.02$ $\mathcal{R}_{\mathrm{Jup}}$, discovered and discussed in detail by Carmichael et al. (2020). Oddly, the primary is metal-rich, which should result in a larger, not smaller, radius for the brown dwarf.

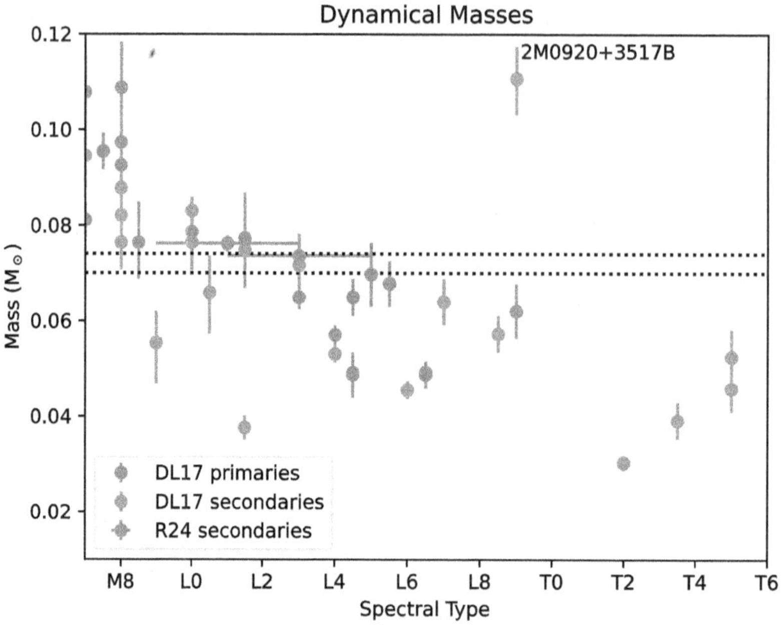

Figure 8.3. Measured individual masses for nearby M, L, and T dwarf binaries from Dupuy & Liu (2017). The primaries are in blue, and the secondaries are in green. Also shown are two brown dwarf companions to more massive stars from Rickman et al. (2024), where the spectral type is estimated from the photometry. The horizontal dashed lines mark plausible solar-metallicity hydrogen-burning limits of 0.070 $\mathcal{M}_\odot$ and 0.074 $\mathcal{M}_\odot$. There are L3 to L5.5 dwarfs whose masses may be consistent with being stars and others of the same type that are definitely brown dwarfs. If 2M0920+3517B is itself an unresolved double, then its large mass can be explained in the framework of traditional models.

dwarf stars: Although the orbits have periods of 20–50 years, precise masses can still be determined from radial velocities and direct imaging. Both studies are consistent with the view that many L0–L2 dwarfs are above the theoretical hydrogen-burning limit, and there are L3 to L5.5 dwarfs whose masses may be consistent with being stars. Here, another major prediction of the theory of very-low-mass stars and brown dwarfs seem to be confirmed with tests.

8.2 Final Thoughts

This book has to this point served as an overview of brown dwarf observations and theory to prepare us to investigate new research questions—and improve on the old answers. There is no question that the unprecedented capabilities of JWST are revolutionizing the field. Indeed, we can be confident that any 2024 compilation of T dwarf and Y dwarf properties will be completely obsolete by 2026. That said, there are huge amounts of data available from other telescopes that can still be exploited. JWST is unparalleled for the study of individual objects but there will always be a role for discoveries from wide-field surveys. We have seen how existing sky surveys have enabled the discovery of most brown dwarfs within 20 parsecs and many more to greater distances. Future sky surveys should extend these distance limits—at least for L and T dwarfs—and greatly improve the population statistics plotted in Chapter 7. On the other hand, there should be low mass objects ($\mathcal{M} < 0.010\ \mathcal{M}_\odot$) with $T_{\text{eff}} < 300$ K. Our prospects for discovering enough of these objects to measure meaningful population statistics is very poor. Leggett et al. (2019) discuss what would be required in a future mid-infrared survey mission to discover these planetary mass objects and probe the population of ejected planets.

Many of the most important questions for the field remain unanswered, especially for the lowest mass objects. The minimum mass of star formation is below the deuterium-burning limit, but what exactly it is, and the corresponding mass function and even formation scenarios for the lowest mass brown dwarfs and planetary mass objects remains uncertain. The L/T transition between cloudy, CO-dominated atmospheres and clear, CH_4 dominated atmospheres had many unexpected properties. What will be revealed for Y dwarfs and cooler objects with water clouds? What lessons will they hold for exoplanets, and how will directly imaged warm or cool gas giant planets compare to brown dwarfs? Even for L and T dwarfs, there is much to be learned. Our models and observations of atmospheric chemistry and condensates clouds have successes but can hardly be considered complete. We have very small samples of metal-poor brown dwarfs. We have not touched on magnetic activity in this book, though there are fascinating observations, and little to nothing is known about rocky planetary companions to brown dwarfs. Review articles and your own research are the next step in becoming an expert on brown dwarfs.

References

Burgasser, A. J., Kirkpatrick, J. D., Cutri, R. M., et al. 2000, ApJL, 531, L57

Carmichael, T. W. 2023, MNRAS, 519, 5177

Carmichael, T. W., Quinn, S. N., & Mustill, A. E. J. 2020, AJ, 160, 53

Dupuy, T. J., & Liu, M. C. 2017, ApJS, 231, 15

Kitzmann, D., Heng, K., Oreshenko, M., et al. 2020, ApJ, 890, 174

Leggett, S., Apai, D., Burgasser, A., et al. 2019, BAAS, 51, 95

Line, M. R., Marley, M. S., Liu, M. C., et al. 2017, ApJ, 848, 83

Luck, R. E. 2018, AJ, 155, 111

Marley, M. S., Saumon, D., Visscher, C., et al. 2021, ApJ, 920, 85

Montes, D., González-Peinado, R., Tabernero, H. M., et al. 2018, MNRAS, 479, 1332

Phillips, M. W., Liu, M. C., & Zhang, Z. 2024, ApJ, 961, 210

Rickman, E. L., Ceva, W., Matthews, E. C., et al. 2024, A&A, 684, A88

Whiteford, N., Glasse, A., Chubb, K. L., et al. 2023, MNRAS, 525, 1375

Zalesky, J. A., Saboi, K., Line, M. R., et al. 2022, ApJ, 936, 44

www.ingramcontent.com/pod-product-compliance
Lightning Source LLC
Chambersburg PA
CBHW082103210326
41599CB00033B/6565